Introduction to Telephones & Telephone Systems

Second Edition

For a complete listing of the *Artech House Telecommunications Library*, turn to the back of this book . . .

Introduction to Telephones & Telephone Systems

Second Edition

A. Michael Noll
Annenberg School for Communication
University of Southern California

Artech House
Boston • London

Library of Congress Cataloging-in-Publication Data

Noll, A. Michael.
   Introduction to telephones and telephone systems / A. Michael
Noll. -- 2nd ed.
      p.   cm.
   Includes bibliographical references and index.
   ISBN 0-89006-550-0
   1. Telephone. 2. Telephone systems. I. Title.
TK6162.N65   1991   91-12112
621.385--dc20   CIP

British Library Cataloguing in Publication Data

Noll, A. Michael.
   Introduction to telephones and telephone systems.-2nd. ed.
   1. Telephony
   I. Title.
   621.385

   ISBN 0-89006-550-0

© 1991 Artech House, Inc.
685 Canton Street
Norwood, MA 02062

International Standard Book Number: 0-89006-550-0
Library of Congress Catalog Card Number: 91-12112

10   9   8   7   6   5

This book is dedicated to the memory of my parents,

Adrian M. Noll

(1902–1965)

Anne R. Noll

(1906–1985)

CONTENTS

PREFACE TO THE FIRST EDITION

The telephone surely must be considered one of the most marvelous inventions of the communication era. Physical distance is conquered instantly, and any telephone in the world can be reached through a vast communication network that spans oceans and continents. The form of communication is both natural and unique, namely, human speech.

Telephone communication can bring joy or convey sorrow. Simply by lifting the telephone handset and dialing a few numbers, we can "reach out and touch someone," as promoted in the television commercial. From its earliest days, the telephone was an instant success. There is no reason to doubt that this success will be diminished now or in the foreseeable future.

The technology of telephone communication is quite simple and straightforward, but the system that makes telephony possible and affordable is quite vast, involving a variety of technologies that have appeared over the telephone's hundred-year lifetime. It is a complex system in its size and scope, yet the basic principles of the technology are understandable with a little study.

The size and importance of telephone communication justifies the need for educated managers and laypersons to understand its workings. Furthermore, the recent restructuring of the telephone industry in the United States—an event that seems destined to repeat itself in other countries—makes it even more essential that many people understand the workings of telephone technology.

The material presented in this book is intended for people without any detailed engineering or technical background. Some of the material does assume a knowledge of basic electricity and electronics, though even these concepts are defined in the glossary at the end of the book. The reader who desires a deeper knowledge of basic communication electronics is referred to the author's *Introduction to Telecommunication Electronics* (Artech House, 1987). However, most of the material should be very understandable even without any knowledge of electricity and electronics. The terms and concepts essential to understanding the material are explained in the text. If they are not, then the reader can usually read on without any loss of understanding of the broader concepts.

The material in this book is based on graduate-level courses taught by the author at the Annenberg School for Communication at the University of Southern California and also at the Interactive Telecommunications Program at New York University.

Four major areas combine to make telephone communication possible. The first area deals with the telephone instrument itself and is called *station apparatus.* The second area deals with the various *transmission media* that carry communication signals over long and short distances. The third area is involved with the techniques and technology for *switching* communication circuits and paths so that any desired telephone can be reached. The fourth area, *signaling,* involves the methods for controlling the operation of the telephone network. Each of these four areas is treated separately in the following chapters. We should always remember that all of these areas work together as a single system to create a telephone communication network.

The systemic aspect of telephone communication is quite apparent in the newest telephone communication service—mobile cellular telephone service. Here, mobile units work with a multitude of base stations transmitting two-way radio signals. The operation of the whole system is controlled by a centrally located computer, which also controls access to the conventional public switched telephone network.

Digital switching and transmission technologies have had a great impact on telephone communication. The digitalization of voice telephony would appear to facilitate a natural migration to an all-digital communication system in which voice telephony and data communication are combined together. Thus, a small amount of material on the basic principles of data networks is included in the book to present a total system perspective of telephone communication.

For nearly a century, telephone communication in the United States was synonymous with AT&T and the Bell System. However, the advent of competition in telecommunication and the restructuring of the Bell System have revolutionized the telecommunication industry. To give an historical perspective to these events, the history of the Bell System is reviewed. An overview, including some personal perspectives on these events, is presented also. Technology has greatly shaped telecommunication, but the fragmentation of the Bell System had a far more cataclysmic effect than any effect ever caused by revolutions in technology.

Dr. Martin C. J. Elton of NYU's Interactive Telecommunications Program encouraged me to try my hand at teaching, and this gave me the opportunity to teach the course that produced the course notes that produced this book. Dean Peter Clarke of the Annenberg School for Communication at USC created an environment to support such scholarly endeavors as writing this book. I appreciate their contributions. In addition, I wish to acknowledge the influence during my early career at Bell Labs of Dr. John R. Pierce, Dr. Edward E. David, Jr., Dr. Peter B.

Denes, Dr. Manfred R. Schroeder, and Dr. Max V. Mathews. They all helped me appreciate the importance of explaining science and technology to nontechnical people in an understandable fashion.

A. Michael Noll
Los Angeles
March 1985

PREFACE TO THE SECOND EDITION

In the five short years since the first edition of this book appeared, a lot has happened to the technology of telecommunication. All new switching machines are digital. Optical fiber forms the backbone of nearly all long-distance communication networks. Cellular telephone service has experienced phenomenal growth. Analog frequency-division multiplexing has been completely replaced by digital time-division multiplexing. Facsimile machines have become as ubiquitous as the telephone in most businesses. This rapid pace of technology made the first edition of this book obsolete and created the need for a revised and updated second edition. Much of the original material has remained, and new material has been added where appropriate.

The writing of this second edition was delayed for two years while I worked with Dr. John R. Pierce in revising his book *Signals: The Science of Telecommunication* (Scientific American Library, 1990). This project gave me an opportunity to reflect on my work with such telecommunication services as picturephone, video teleconferencing, and videotex. These reflections became the basis for a new chapter on services in this second edition.

Karen Ann Kurlander and Steve Fleischer at Bell Atlantic Mobile Systems reviewed the chapter on cellular telephone service. Various sections of the switching chapter were reviewed by Peter Settles and Carl Griffith of Northern Telecom and by Richard Q. Hofacker, Jr., and his colleagues at AT&T Bell Labs. I thank all of these people for their help and for correcting many of my mistakes and misunderstandings. I also thank my editor, Dennis Ricci, for his assistance in producing both editions of this book. This second edition was written during my sabbatical from the Annenberg School for Communication, and I appreciate the support of the School during this project.

The writing of this second edition created an opportunity to improve on the quality of the figures; they were redrawn from my hand drawings by Richard Cook at the Annenberg School using a computer. My publisher also worked on improving the overall appearance of the book. We hope that you are pleased with the result of our efforts.

A. Michael Noll
Stirling, New Jersey
January 1991

Chapter 1
INTRODUCTION

THE KEY TO THE INFORMATION AGE

Since the 1970s, communication technology has made great advances, offering capabilities that were unthinkable a decade earlier. The bandwidths of the new transmission technologies are vast and virtually unlimited, offering the ability to carry large numbers of voice, data, and video signals. New electronic switching systems are reliable and small in size, compared to the electromechanical systems of the past, and the computers that control them offer great flexibility in terms of new programmable services. Simultaneously with these technological advances, the whole telecommunication industry has undergone dramatic restructuring.

It is easy to be swept away with an uncontrolled enthusiasm for the new communication technologies and the many new services they could make possible. An optical fiber to every home could provide a tremendous variety of television programs on demand. Videotex could bring all the world's information to every home in the form of information frames displayed on the home computer or television set. The capabilities of the new transmission and switching technologies could make two-way video telephones feasible. The home itself could be controlled by a home computer that could be accessed remotely for monitoring purposes. Banking could be conducted electronically by telecommunication. Many employees could work at home using telecommunication technologies to telecommute to meetings. Business meetings with geographically dispersed groups of people could be conducted by two-way video teleconferencing, thereby eliminating the need for physical travel. Television could become three-dimensional in high definition on large, thin displays that would fill the living room with images and sound. Voice, image, data, and video communication could all come together along with computers and telecommunication to create a single, integrated system. And all this could be only the beginning!

This utopian view of communication technology and all it could do is cloaked in a sense of euphoria that makes it difficult to separate hyperbole from reality. Videotex promises to provide instant access to computerized data banks, but consumer acceptance of it has been very disappointing. Two-way video teleconferencing continues to excite much interest, but the idea is quite old now and its actual use is still minor. Electronic mail and electronic bulletin board services attract much attention, but many users seem to be computer hobbyists and no one seems to know how to develop a strategy to enter this market in a profitable way. The picturephone was tried decades ago and failed; most consumers simply do not want to be seen while speaking on the telephone.

Yet, there is one communication service that is over 100 years old and attracts little euphoria or excitement. We simply take it for granted that it is available in nearly every home and workplace in the nation and in the world. We use this service eight times per day, on the average. With this service, emergency aid can be quickly summoned, airline reservations made, goods ordered and purchased, and funds transferred. No special skills are needed to use this service. This ubiquitous service that totally penetrates our lives is the *telephone*. Telephone service is truly synonymous with communication, but is so taken for granted that we frequently forget its paramount position in our lives and society.

Telecommunication is the key component of the information age and the services that form the basis of our information economy, and the data clearly support this claim. Telecommunication revenue in 1987 was $154 billion—far more than the $50 billion for outlays for computers; $61 billion for newpapers, books, and periodicals; and $53.5 billion for the television industry, including set sales, advertising receipts, videotape, and CATV. In terms of telecommunication traffic, there is no doubt that voice telephone service is the major contributor, although data traffic is an important aspect of computerized information services and cannot be neglected. The main thrust of this book is the telephone and the public switched network and the technology that makes telephone service possible.

A SYSTEMS PERSPECTIVE

The most visible component of telephone service, or *telephony,* is the *telephone instrument* itself. The telephone instruments in the home or business are all connected by wires that are referred to as *intrapremises wiring.* The telephone instruments themselves, along with such other devices as facsimile machines and telephone answering machines, are referred to as *customer-premises equipment,* or CPE. As shown in Figure 1.1, the intrapremises wiring is brought together within the customer's premises at some central point. This central point then connects with a single pair of wires, called the *local loop,* which connects the premises with a distant central location called the *central office.* The intrapremises wiring and CPE are protected from dangerous voltages, such as those caused by lightning, by a fuse called the *protector block,* which is located at the point of connection of the

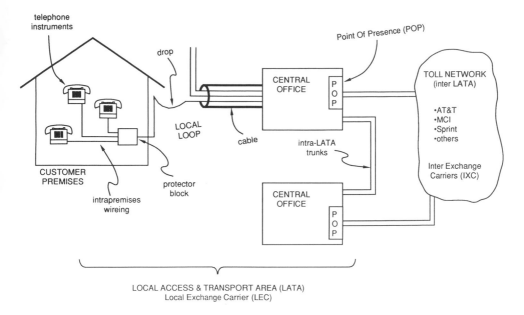

Figure 1.1 A telephone call involves much switching and transmission equipment "beyond the dial." Each telephone is connected to a central office by a pair of wires called the local loop. The first stage of switching occurs at the serving central office. Calls to another office within the local access and transport area are carried over interoffice trunks. Calls outside the LATA are handled by interexchange carriers over their own transmission and switching facilities. The point of presence is the place where the IXC connects to the facilities of the local exchange carrier.

intrapremises wiring to the local loop. The local loop is so called because there are two wires, which during the early days of telephony were visualized as forming a loop connection.

The wire pairs of many local loops from many premises are combined together in a large *cable*. The cable might be fastened to telephone poles or buried underground. All the cables from a local area terminate at the central office that serves all the premises within its jurisdiction. Equipment in the central office supplies many useful functions. The central-office equipment supplies dial tone to signal the telephone user that the equipment is ready for dialing to commence. The equipment receives the dialed number, interprets it, and begins the steps necessary to make the connection over which the telephone conversation will occur.

If the called party is served by the same central office, the connection is made immediately and most simply. If the called party is located within the same immediate geographical area but is served by another central office, connection is first made over communication circuits, called *trunks*, between the central offices and then the final connection is made at the central office serving the called party.

If the called party is located far away in a distant part of the state or in another state, then the equipment in the calling party's central office will make a connection to the *long-distance network*. Equipment and facilities in the long-distance network will create a connected circuit to the distant central office serving the called party, where the final connection will occur. All this typically happens within a few seconds.

The local portion of the telephone system is called a *local access and transport area* (LATA). Calls between LATAs are referred to as inter-LATA calls. Local telephone companies are restricted to providing service within their LATAs (intra-LATA service), while long-distance companies provide inter-LATA service. The local telephone companies are sometimes called *local exchange carriers* (LECs), and the long-distance companies are called *interexchange carriers* (IXCs). The place where the IXC connects to the LEC is referred to as the point of presence, or POP. Some large users may connect directly to an interexchange carrier, thereby avoiding the local exchange carrier; this is called *bypass.*

During the early days of telephony, and today too, a flat rate was charged, which allowed the calling party to make unlimited calls within a local area. Calls outside that area were billed on the basis of the distance and the length of the call. These long-distance calls became known as *toll calls,* and the long-distance network was called the *toll network.* The telephone user subscribed to telephone service on a monthly basis, just as today, and became known as the telephone *subscriber.* The nation's toll network used to be operated solely by the Long Lines Division of the American Telephone and Telegraph Company (AT&T). Today, a number of firms provide long-distance service, and the nation's long-distance network has become the synthesis of a number of separately operated facilities.

We thus see that there is considerable equipment and facilities "beyond the dial" are necessary when we make a telephone call. The telephone instrument, along with a network of switched transmission circuits, enables any telephone to be connected quickly to any other telephone, anywhere on our planet. This switched network is an extensive facility that allows the public to conduct voice conversations over almost any distance. The public switched network not only connects telephones for voice conversations but can connect a computer terminal to a distant electronic data bank, or a facsimile machine to another facsimile machine to transmit a picture or a letter across oceans. The same network can carry the sound of an ill heart to a specialist many miles away for sophisticated computer analysis and diagnosis. The telephone system and the public switched network are a flexible resource that serves many communication purposes. It is here today available for our use, day in and day out, at our beck and call.

THE VISION

The telephone is generally acknowledged to have been invented by Alexander Graham Bell in 1876 with the assistance of his colleague Thomas A. Watson. Bell had the vision to see a world wired together by his invention. After a decade or so of intense competition, the Bell companies emerged as the monopoly supplier of telephone service in the United States. Theodore N. Vail founded the Bell System as a vertically integrated monopoly to supply telephone service. Now, nearly a century later, as a result of the drastic restructuring of the Bell System that occurred on January 1, 1984, competition has returned to the provision of telephone service. In many ways, technology led to the breakup of the Bell System.

REGULATION

The provision of telephone service is subject to regulation on a number of governmental levels. Interstate long-distance service is regulated by the Federal Communications Commission. Intrastate toll service and local service are regulated by state public utility commissions, and possibly county and city commissions in some circumstances. As a result of the divestiture of the Bell local telephone companies from AT&T, the US District Court, the US Congress, and the US Department of Justice now also have a voice in telecommunication policy and regulation. From a policy perspective, telecommunication has many players and can appear very confusing and contradictory at times. A later chapter attempts to give some perspective on divestiture and telecommunication policy.

THE CONSUMER SHAPES THE FUTURE

Telephone communication has not stood still during its century of existence, and new services and new uses of the public switched network continue to emerge. Some of these new services, such as video teleconferencing and videotex, are technologically most impressive, but consumers have expressed little interest in many of them. Other new services, such as mobile cellular telephony, are also technologically impressive, and consumers have responded enthusiastically to them. Technology indeed suggests many new services, but ultimately it is the consumer who determines success and failure in the marketplace. Later in this book, we will examine many of these new services and the apparent reasons for their success and failure.

SYSTEM ELEMENTS

The technology of a telephone communication system consists of four major system elements. As we have already mentioned, the first element is the telephone instrument itself, along with any other form of apparatus that is attached at the telephone station. The second element, transmission, is involved with the various media used to transmit a telephone signal over distance. The third element, switching, concerns the various ways in which one telephone circuit is connected to another telephone circuit. The last element, signaling, is involved with how the telephone network is controlled and instructed to make an actual connection. The technical content of this book thus is organized into four main chapters: station apparatus, transmission, switching, and signaling.

Chapter 2
STATION APPARATUS

INTRODUCTION

The most visible aspect of telephone service is the telephone instrument itself. However, as we shall see much later in this book, there is considerably more "beyond the dial" to telephone service. However, what could be more natural than examining the workings of the telephone instrument itself to begin our understanding of the workings of telephone service?

The telephone is only one type of apparatus that can be stationed on a customer's premises. Telephone answering machines and facsimile units are other types of *station apparatus*. Telephones and other station apparatus are considered customer premises equipment, or CPE.

BELL'S INVENTION

March 10, 1876, is a very important date in the history of telephony for it was on that day that the first working model of a telephone was demonstrated by Alexander Graham Bell and his assistant, Thomas A. Watson. The patent application for the telephone had been filed a month earlier, on February 14, 1876, by Bell. However, a disclosure for a telephone had also been filed that same day by another inventor, Elisha Gray. Although Bell's invention was upheld in a split decision by the US Supreme Court, mystery and controversy continue to cloud the question of whether Bell had learned of and copied some of the key features in Gray's disclosure.

The key feature in the telephone demonstrated by Bell was the use of a variable-resistance transmitter, which generated an electric current that varied enough to generate an ample acoustic signal at the receiver. The idea of a variable-resistance transmitter was a key feature in the caveat, or warning to other inventors, filed by Elisha Gray on that fateful day of February 14, 1876. Bell's patent filing also disclosed the idea of a variable-resistance transmitter, but the disclosure was in the form of a handwritten marginal notation, almost as if it were an afterthought or last-minute addition. We will never know whether Bell, possibly through his supporters, had somehow learned of Gray's idea and added the marginal notation later or whether all this was simply pure coincidence. But history certainly credits Bell with the invention of the telephone, and there is no doubt that Bell had the vision to foresee the vast business and social value of universal telephone service.

The telephone as invented and demonstrated by Bell consisted of a microphone, called a *transmitter,* and a small loudspeaker-like device, called a *receiver,* connected by a pair of wires and in series with a battery. The series battery was essential to the concept of a variable-resistance microphone, as we shall soon see. Figure 2.1 shows a model of Bell's first telephone.

Figure 2.1 Photo of a model of the first electric telephone's liquid transmitter (on the left) and tuned-reed receiver (on the right). (Courtesy of AT&T Bell Labs.)

Bell's initial idea for a telephone system used an electromagnetic transmitter that was, in essence, a small loudspeaker-like device hooked up in reverse. The problem was that such an electromagnetic transmitter could not generate enough power to produce much sound at the receiver. The solution to this problem was the variable-resistance transmitter in which the sound wave did not directly generate power but rather controlled the power flowing in an electric circuit.

As shown in Figure 2.2, the variable-resistance transmitter, the receiver, and a battery are all connected in series. The battery causes a constant electric current to flow through the circuit. That current I, according to Ohm's law, is the electromotive force E of the battery divided by the total resistance R in the electric circuit,

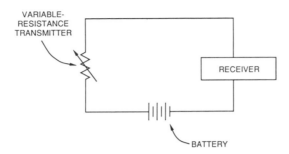

Figure 2.2 The variable-resistance transmitter was key to the working of the early telephone. The sound wave did not generate the power necessary to generate sound at the receiver but rather controlled the electric power flowing in the circuit. An external battery supplied the electric power.

or $I = E/R$. The total resistance comes primarily from the resistance of the microphone since the receiver is a coil of wire, which has very little resistance. When the telephone user speaks into the transmitter, the increases and decreases in the sound pressure cause the resistance of the transmitter to likewise increase and decrease. These variations in resistance cause the current to decrease and increase, and the resulting variations in current are quite large and more than sufficicent to generate a large acoustic output from the receiver.

The real problem for Bell was how to produce a variable-resistance transmitter. His solution, used in that first demonstration on March 10, was the liquid transmitter. Sound waves were focused on a diaphragm to which a small wire was attached. The wire was immersed in a small metal cup containing a mixture of water and acid (Figure 2.3). The wire moved up and down in the liquid in response

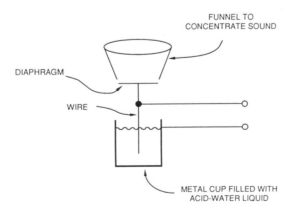

Figure 2.3 Bell's liquid transmitter consisted of a metal cup filled with a mixture of water and acid. The sound wave caused a small wire to move up and down in the liquid. The electric resistance between the wire and the cup thus varied in response to the sound.

to the sound waves, and as it did so, the electric resistance between the wire and metal cup decreased and increased. A battery supplied a constant voltage to the circuit, and it—along with the varying resistance of the transmitter—produced the varying current necessary to drive the receiver.

Bell's first receiver was a reed tuned to vibrate in response to the varying current. Later, the reed was replaced by a coil of wire wound around a permanent magnet, as shown in Figure 2.4. The varying current in the coil of wire created a varying electromagnetic field, which attracted a metal diaphragm. This electromagnetic receiver was invented in 1876 by Watson, and the same principle continues to be used in today's telephone receivers.

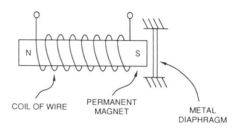

COIL OF WIRE   PERMANENT MAGNET   METAL DIAPHRAGM

Figure 2.4 Watson's electromagnetic receiver was somewhat like a small loudspeaker. A coil of wire wound around a permanent magnet generated a varying magnetic field, which created a varying attraction to a metal diaphragm.

TRANSMITTERS

During the early days of telephony, the key challenge was obtaining a signal of sufficient level so that the person at the other end of the connection could hear the speech. Because electronic amplification was not known then, the solution was to obtain as strong a signal as possible from the transmitter. Telephone instruments producing a larger output signal than others were judged superior, and thus the telephone company supplying those instruments had an advantage over its competitors.

The acid in Bell's liquid transmitter made it impractical for consumer use. Bell also investigated an electromagnetic transmitter in which a small magnet was attached to a diaphragm and the speech caused the magnet to move near a coil of wire, thereby inducing an electric signal. However, the output from this type of magneto transmitter was far too small.

In 1877, the "wizard of Menlo Park," Thomas Alva Edison, invented a variable-resistance transmitter utilizing a small disk of carbon in the form of compressed lampblack. The Edison transmitter produced a high output, and the non-

Bell companies that used it had an advantage over the Bell companies. However, the most important principle in developing a high-output transmitter was demonstrated in 1878 by David E. Hughes. Hughes demonstrated that one nail laid across two other nails, between which an electric circuit was connected as shown in Figure 2.5, would produce large variations in resistance when the loose nail was vibrated by an acoustic signal. He also demonstrated that such *loose contacts* with carbon would produce the same effect, and he revived the term *microphone* to describe a transmitter device based on the loose-contact principle.

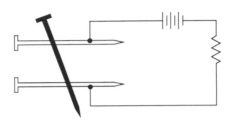

Figure 2.5 In 1878, David E. Hughes demonstrated that one nail laid loosely across two others would vibrate in response to an acoustic signal. That vibration would produce a varying resistance in the electric circuit formed across the two stationary nails. He revived the term *microphone* to refer to transmitters based on this loose-contact principle.

A number of inventors investigated loose-contact microphones. Emile Berliner was one of them. Another was Francis Blake, who had joined the Bell company in 1878. Blake's loose-contact transmitter, invented that year, consisted of a platinum bead, held by a light spring, which vibrated against a hard carbon disk. The Blake transmitter then became the standard transmitter in Bell telephones. But more innovations in transmitters were soon to come.

On September 16, 1878, Henry Hummings received a British patent for a telephone transmitter using a cavity partially filled with finely divided carbon. The electric circuit was completed across the cavity. The American Bell Telephone Company purchased the rights to Hummings's invention and in 1885 redesigned it to eliminate some of the packing problems that occurred over time with the carbon. In 1886, Thomas Alva Edison improved Hummings's invention through the use of roasted granules of anthracite coal in the cavity, as depicted in Figure 2.6. The many granules create many paths for the electric current, and hence the dc current supplied from the battery can be much higher than with previous designs. The diaphragm causes the carbon granules to vibrate against each other thereby causing the resistance to vary.

The carbon-granule transmitter was a major innovation in telephony and is still in use today, though improved in characteristics and manufacturability (see

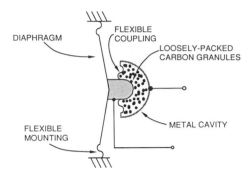

Figure 2.6 The carbon transmitter has a small metal cavity partially filled with carbon granules. The many granules are all in contact with each other and create many electric paths across the cavity. Since there are many paths, the current across the contacts can be higher.

Figure 2.7). The carbon transmitter, or microphone, does not respond well to small acoustic signals and thus is good at discriminating against acoustic background noise. Furthermore, if the speech signal is very strong, the carbon microphone overloads somewhat, thereby protecting the telephone connection from excessive signal levels. These characteristics plus its low cost have accounted for the long life of this technology. The carbon granules can stick or pack together, particularly in wall-mounted handsets, although a simple solution is to slam down the handset to loosen the granules.

EARLY RINGERS AND DIALERS

Bell quickly realized that the person making a telephone call needed a way to signal the other person to pick up the telephone. People tapped their transmitters, but the signal at the other end was too weak to be heard. Bells and buzzers of various kinds were tried, but were not loud enough. Thomas Watson discovered the ultimate solution: a ringer consisting of a polarized call bell, shown in Figure 2.8. A permanent magnet supplied the energy to strike the bell hard enough to produce a load sound. A pivoting metal armature was made to move by a varying electric current generated by a magneto in the calling telephone and operated by a hand crank. A hammer was attached to the armature, and it moved to and fro striking two bells quite loudly. The patent application for Watson's ringer was filed on August 1, 1878. Today's telephone ringers are essentially the same as the one invented by Watson.

Early telephone ringers were connected in series with the line and hence caused a transmission loss because the voice signal had to flow through the ringer

Figure 2.7 Photograph of a modern carbon-granule transmitter showing the cavity and the carbon granules. (Courtesy of AT&T Bell Labs.)

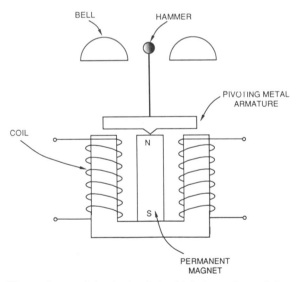

Figure 2.8 Thomas Watson invented the dual-polarized telephone ringer. A hammer is attached to a pivoting metal armature. The use of a permanent magnet increases the stength of the strikes of the hammer against the two bells. The two coils that carry the varying electric ringing signal are connected in series.

too. In 1890, a Bell employee, John J. Carty, invented a ringer with a high impedance of 1000 ohms. *Impedance* is the opposition to the flow of alternating current
in an electric circuit. Since Carty's ringer had a high impedance, it could be connected in parallel across the line and for voice frequencies would not draw appre-
ciable current from the telephone or from the line, as shown in Figure 2.9. The
high-impedance ringer required a higher voltage to operate it; an ac signal of 75
volts rms was used then and ever since. A capacitor is connected in series with the
two ringer coils to prevent dc from flowing through the ringer.

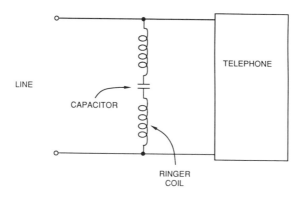

Figure 2.9 The telephone ringer is connected across the telephone line and in parallel with the rest of
the telephone instrument. A capacitor is inserted in series with the two coils to block dc from
flowing through them.

At first, telephones were installed as private-line systems connecting only two
instruments, as shown in Figure 2.10. In early 1878, the first telephone exchange
was formally opened. It allowed 21 customers to reach each other by means of a
central switchboard. The manual switchboard was operated by young men, and
later by more polite and courteous women. An early switchboard is shown in Figure 2.11.

The first automated switchboard, installed in 1892, was invented by Almon
B. Strowger. Strowger's automated switchboard consisted of an electromechanical
switch that was operated by pulses from the telephone instrument. These pulses
were produced when the user rapidly pushed a button on the telephone.

Strowger and his associates formed their own company, the Automatic Electric Company, to manufacture and sell automatic telephone systems. These systems
were purchased and used by the non-Bell, or independent, telephone companies. It
was not until 1919 that the first Bell exchange was equipped with Strowger's automatic switching system. In 1896, A. E. Keith, J. Erickson, and C. J. Erickson, all

PRIVATE-LINE SERVICE:

EXCHANGE SERVICE:

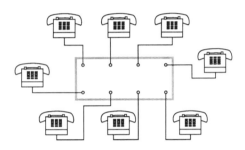

Figure 2.10 Private-line service simply connected together two telephones and no other telephones could be reached. With exchange service, any other telephone served by the exchange could be reached. Human operators made the cross connections from one telephone line to another.

associates of Strowger, invented a rotating fingerwheel for dialing the desired telephone number. Telephone dialing is explained later in this chapter.

THE LOCAL LOOP

During the early days of telephony, a length of metallic wire connected the telephones and the earth itself was used as a source and as a sink for the flow of electrons that comprises electricity, as shown in Figure 2.12. Since nearly everyone, including the power companies, used the ground for electric circuits, considerable noise and interference resulted. The solution was devised by John J. Carty in 1881. He used a second metallic wire to complete the electric circuit, thereby avoiding the ground path, as shown in Figure 2.13. By slightly twisting together the two wires, noise from airborne electromagnetic interference was also reduced.

The two wires formed a loop between the telephone subscriber's premises and the local office of the telephone exchange. Thus, the wires were called the *local loop*. The term is still in use today to refer to the circuit between the subscriber's premises and the local office.

Figure 2.11 Photograph of operators at work at a pyramid switchboard in Richmond, Virginia, in 1881. (Photo courtesy of AT&T Bell Labs).

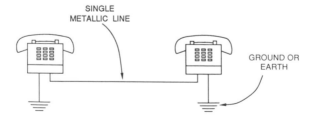

Figure 2.12 The use of the earth itself as a ground return for the flow of electrons that forms electricity created a source of considerable electric interference.

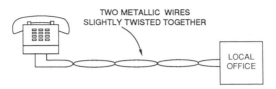

Figure 2.13 The use of a second metallic wire to form an electric circuit was suggested in 1881 by John J. Carty. The circuit between the telephone at the customer's premises and the local office is called the local loop. The two wires are slightly twisted together to minimize the effects of airborne electromagnetic interference.

POWER AND RINGING

A telephone is powered by *direct current* (dc). Early telephone instruments included their own internal battery to supply this direct current. In 1894, a common battery at the central exchange was used for the first time to power all the telephone instruments connected to the exchange. This common battery had a voltage of 48 volts. Common battery at 48 volts has been used ever since.

When a telephone handset is on its cradle and the telephone is not in use, the telephone instrument does not draw any direct current from the central office. When service is desired, the user lifts the handset from its cradle, and a dc electric circuit is completed so that the telephone instrument draws dc over the local loop. The initiation of this flow of dc is sensed by equipment at the central office so that the party requesting service can be identified and served, as shown in Figure 2.14.

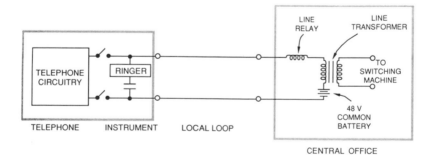

Figure 2.14 The circuitry in a telephone instrument that operates the instrument draws direct current from the local loop and is connected by the switch hook. The flow of dc over the local loop is sensed by a line relay at the central office. The 48-volt common battery is also located at the central office. A transformer connects the local loop to the switching equipment so that only the ac speech signal continues. The ringer in the telephone instrument is always connected across the line, and a capacitor prevents direct current from flowing through it.

In early telephone instruments, the receiver was hung on a hook that activated a switch to control the flow of direct current to the instrument. This hook was called the *switch-hook*. The mechanism in today's telephone instrument that controls the flow of direct current is still called the switch-hook. Figure 2.14 shows the connection of the switch-hook in a telephone instrument.

The telephone ringer is always connected across the telephone line. A capacitor in series with the ringer prevents direct current from flowing through it. The ringer has a high impedance, or opposition to the flow of alternating current, and thus little current from the speech signal flows through the ringer. The ringer voltage of 75 volts rms appears across the ringer and causes it to ring. The frequency of the ringer voltage is 20 Hz. The ringer voltage is on for 2 seconds and is then off for 4 seconds, as shown in Figure 2.15. When the called party answers the telephone by lifting the handset from its cradle, the switch-hook closes, direct current flows through the instrument, and central office equipment sensing that flow knows that the party has answered and stops the ringing signal.

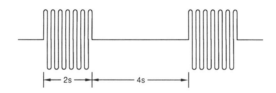

Figure 2.15 The ringing voltage consists of bursts of a pure tone, or sine wave, at a frequency of 20 Hz and with an rms electromotive force of 75 volts. The bursts are on for 2 seconds and off for 4 seconds.

INDUCTION COIL

The telephone first demonstrated by Bell was a one-way system. Telephone service was made two-way through the use of identical transmitter-receiver combinations in each instrument. The transmitter and receiver at each instrument were connected in series with each other and also with the battery that powered the transmitter, as shown in Figure 2.16. This arrangment had some serious problems.

The receiver typically had a much higher impedance than the transmitter. This greatly reduced the current flow in the circuit and, consequently, greatly reduced the received speech signal. The direct current from the battery flowed through the receiver and could oppose the magnetic flux of the permanent magnet in the receiver, thereby reducing the efficiency of the receiver. The transmitter typically had an impedance of about 5 ohms (Ω) which did not match the impedance of the line, which was a few hundred ohms. This impedance imbalance reduced the

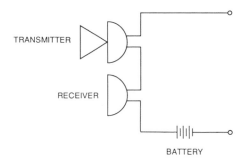

Figure 2.16 The transmitter, receiver, and battery were all connected together in series in early tele-phone instruments. Because the impedance of the transmitter was more than the receiver, the current in the circuit was reduced.

energy transfer from the transmitter to the line. The solution to all these problems was the use of a *transformer,* initially called a *repeating coil* and later an *induction coil,* which was patented for this use in 1878 both by Berliner and by Edison.

As indicated in Figure 2.17, the transmitter and battery were connected in series with the primary of the induction coil. The secondary of the induction coil, the receiver, and the line were connected in series. An induction coil responds only to alternating current, and hence the direct current from the battery affected only the transmitter. The number of turns in the primary winding of the induction coil

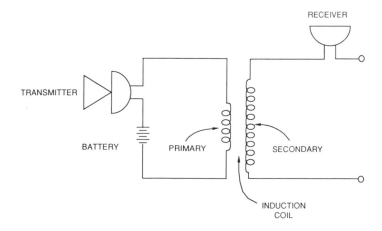

Figure 2.17 The induction coil, or transformer, was used to separate electrically the transmitter and receiver in a telephone instrument. The number of turns in the secondary was greater than in the primary so that the transmitter voltage was stepped-up before being sent down the line.

was about 5 to 10 times the number of turns in the secondary winding. This is called a *step-up transformer,* and it gives an increase in the voltage sent down the line from the transmitter. Furthermore, this type of transformer steps up the low impedance of the transmitter to match better the higher impedance of the line. The induction coil created a large improvement in the quality of early telephones. But an old problem remained—sidetone.

In early telephone instruments, the speech signal created by the transmitter passed through the receiver. Thus, a loud signal of one's own speech issued from the receiver, and this was particularly annoying when the receiver and transmitter were combined together into a single handset. The effect of hearing one's own speech while talking is called *sidetone.* As shown in Figure 2.18, we hear our own

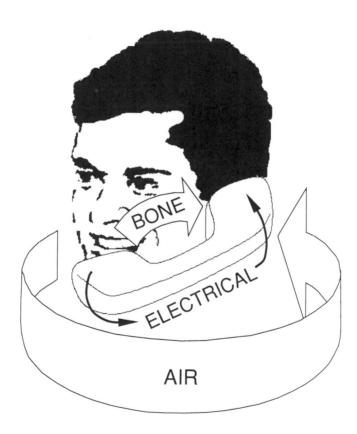

Figure 2.18 While speaking on the telephone, we hear our own speech through air conduction from mouth to ear, through bone conduction, and also through an electrical path from the telephone transmitter to the receiver. The hearing of one's own speech while talking is called sidetone.

speech while talking through air conduction from the mouth to the ear and also through bone conduction. The electrical path from the telephone transmitter to the receiver creates a third path. A certain amount of sidetone is normal, and we use sidetone to adjust our speaking level. However, the amount of sidetone coming from an early telephone instrument was very loud, and users either lowered their speech or moved the telephone handset away. Either way, the transmitter signal was reduced, and the party at the other end received a much smaller signal; this was not an acceptable situation.

The solution was patented in 1918 by G. A. Campbell, an AT&T research engineer. Campbell showed that an induction coil could be connected in such a way as to balance or cancel the electrical sidetone. The actual anti-sidetone circuitry was not used commercially, however, until 1930 because of the need for exhaustive testing as well as the additional cost.

ANTI-SIDETONE CIRCUIT

A center-tapped transformer is the basis of the anti-sidetone circuit, as shown in Figure 2.19. An electrical circuit, called a *balance network,* is needed to match or duplicate the impedance of the telephone line as seen by the telephone instrument. The receiver is connected to the secondary winding of the transformer. The trans-

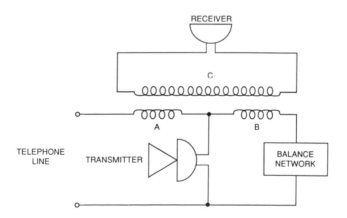

Figure 2.19 A center-tapped transformer is used in the anti-sidetone circuitry invented to reduce electrical sidetone to an acceptable level. The receiver is connected to the secondary winding of the transformer, C. The balance network attempts to duplicate the impedance of the telephone line as seen from the telephone instrument. The current generated by the transmitter divides equally through the primary windings of the transformer, A and B. The effects of the two divided currents cancel in the secondary winding, and no signal is heard from the receiver.

mitter is connected at the center tap of the primary winding. Assuming that the balance network perfectly matches the impedance of the line, the electric current generated by the transmitter will divide equally at the center tap and exactly half will flow in opposite directions through the primary windings. The induced voltages at the secondary winding have opposite polarities and cancel. Thus, no speech signal generated at the transmitter appears at the receiver. A signal coming from the line flows in the same direction through the primary windings in the transformer and induces a voltage at the secondary to activate the receiver.

Local loops vary in their impedances depending on their length and other characteristics. Thus, no fixed balance network can ever perfectly match exactly the impedance of the local loop. Therefore, an imbalance in the current flowing through the primary windings occurs, and some speech leaks to the receiver and is heard. As long as this leakage is not too much, it is acceptable and actually is desirable because some sidetone gives the telephone instrument a sense of being alive. Many people blow into the transmitter to determine that the telephone is working. What they hear as a result is electrical sidetone.

Modern telephone instruments have circuitry in the balance network to compensate for different line lengths, and in this way, improved sidetone balance is obtained. However, some sidetone is desirable, and perfect balance, which could probably be achieved with today's electronic circuitry, is not desirable by reason of human-factor considerations.

The anti-sidetone circuit of current telephones is a little different from early telephones, as can be seen from Figure 2.20. The transmitter and center-tapped primary are the same. However, the receiver is connected with the balance network, and the secondary winding is depicted schematically in such a way that the conventional transformer relationship is not readily apparent.

The local loop is a two-wire circuit. The transmitter and the receiver in the telephone instrument form a four-wire circuit, as shown in Figure 2.21. The induction coil in the anti-sidetone circuitry in the telephone instrument performs this two-wire–to–four-wire conversion. Later in this book, we shall see how similar two-wire–to–four-wire conversion is needed for long-distance telecommunication. This conversion is performed by a transformer quite similar to the induction coil used in the anti-sidetone circuitry, except that it is called a *hybrid coil*. The induction coil in the telephone instrument is sometimes also known as a hybrid coil.

DIALERS

The purpose of dialing is to specify the telephone number of the called party to the switching equipment at the central office. That number can be specified either by *dial pulses* or by *touch-tones*. Dial pulses are produced by the *rotary dialer*. The

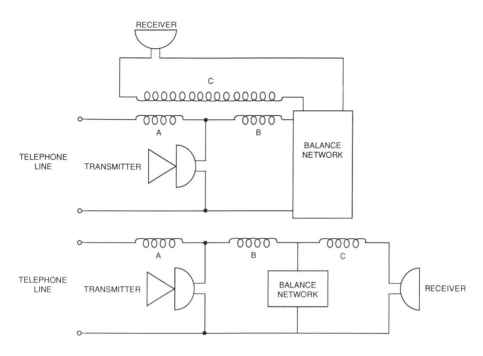

Figure 2.20 In today's telephone instruments, the balance network and the receiver are connected (top). The actual schematic is usually drawn in such a way (bottom) that the transformer relationship of the three windings is not readily apparent.

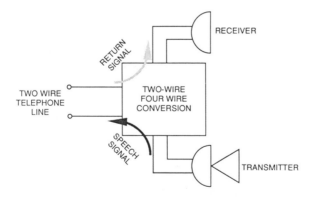

Figure 2.21 The two-wire local loop must be interfaced to the four-wire circuit formed by the transmitter and receiver in the telephone instrument. This two-wire–to–four-wire conversion is performed by the induction coil in the anti-sidetone circuitry.

rotary dialer creates interruptions in the flow of direct current through the telephone instrument, and these interruptions, or pulses, are sensed and counted by the switching equipment at the central office. The rotary dialer is a mechanical device, as shown in Figure 2.22. As it is turned, a spring is wound. After it is released, this spring causes the dial to turn in the oppposite direction, and a small governor controls its rate of rotation. While the dial is rotating during its return, it interrupts the flow of line current, thereby creating dial pulses at a rate of 9 to 11 pulses per second. The line current is interrupted for about 60 percent of the time interval for each dial-pulse cycle. If the local loop is short, as is usually the case for the telephone lines to a private branch exchange (PBX), the dial-pulse rate can be increased to 20 pulses per second.

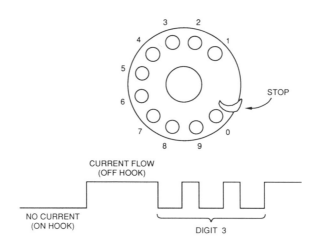

Figure 2.22 The familiar rotary dial generates dial pulses during its return. These dial pulses occur at a rate of 9 to 11 per second. A dial pulse is an interruption in the flow of direct current and lasts 60 percent of each pulse cycle. There is 0.7 second between dialed digits.

Although there are still many rotary-dial telephones in use, most newer telephones use touch-tones with pushbutton dialing. Touch-tone dialing is much faster than dial-pulse dialing. The pushbutton assembly is called a *dial pad.*

Each touch-tone digit is a unique combination of two single-frequency tones, as shown in Figure 2.23. The frequencies are arranged in a matrix. As the button is pushed for a specific digit, the appropriate combination of two tones is generated, corresponding to the intersection of the vertical and horizontal axes. The frequencies corresponding to the horizontal axis are called the *low band* and are 697, 770, 852, and 941 Hz. The frequencies corresponding to the vertical axis are called the

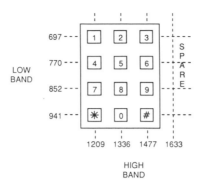

Figure 2.23 The dial pad used for touch-tone dialing. When a button is pushed, two tones at the frequencies corresponding to the intersection of the vertical and horizontal axes are produced. Equipment at the central office senses the frequencies of the tones and determines the dialed digit. A fourth vertical column at a frequency of 1633 is available, but is not yet used.

high band and are 1209, 1336, 1477, and 1633 Hz. For example, if the digit 8 were pushed, two tones would be generated simultaneously and sent over the telephone line: one at 852 Hz and one at 1336 Hz. The received tones are filtered and detected at the central office to determine the dialed digit.

Touch-tone dialing was initiated in 1963, and many telephone companies continue to charge extra for the service, even though touch-tone equipment is standard at the central office. Thus, a consumer might acquire a touch-tone telephone, but it might not work if the consumer has not ordered touch-tone service from the local telephone company. Some newer telephones have a dial pad with pushbuttons, and a switch selects whether dial pulses or touch-tones are created when a button is pushed. These types of telephones work on any line, as do old-fashioned rotary dial phones.

CIRCUIT DIAGRAM

The major elements of a telephone instrument are the ringer, the switch-hook, the dialer, the transmitter, the receiver, and the anti-sidetone circuit (or hybrid). The relationship of these elements is shown in the block diagram of Figure 2.24.

An actual telephone instrument contains very little circuitry compared to the complexity of today's other electronic devices. The circuit diagram for a GTE telephone instrument is shown in Figure 2.25. The ringer is depicted as a coil of wire, and the two lines next to the coil indicate that it has an iron core. The capacitor C_0 in series with the coil prevents direct current from flowing through the coil. The

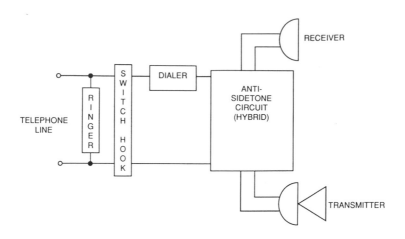

Figure 2.24 The major elements in a telephone instrument.

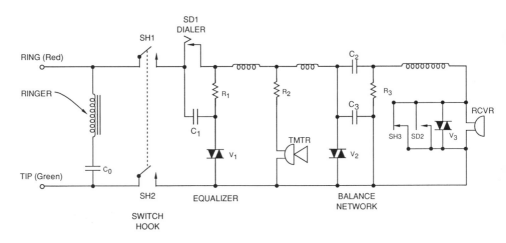

Figure 2.25 Circuit diagram of the GTE Starlite® telephone (type 182).

switch-hook is connected in parallel with the telephone line and is shown as two switches, SH1 and SH2. The switches are open when the handset is on-hook and are closed when the handset is off-hook.

A dial-pulse rotary dialer is shown as a switch, SD1. As the dial is released after being turned, the switch SD1 alternately opens and closes, creating interruptions, or pulses, in the flow of direct current. A capacitor, C_1, is connected across the contacts of the dialer to suppress the creation of sparks that could otherwise interfere with radio and television reception. This capacitor also prevents the ringer from being weakly activated by the dial pulses during dialing, an effect called "tinkling." For this reason, the capacitor along with the resistor R_1 is sometimes called an "anti-tinkling filter."

The three windings of the center-tapped anti-sidetone transformer are shown in a fashion similar to Figure 2.20. The receiver is connected to the secondary of the transformer. Two switches are connected across the receiver. One of them, SH3, is closed when the handset is on-hook. Its purpose is to protect the receiver from any ringer current that might inadvertently leak into the reciver circuit and damage it. This switch opens when the telephone is off-hook and is in use. The second switch, SH2, closes during dialing and prevents the dial pulses from being heard as loud clicks in the receiver.

A *varistor* (or *variable resistor*) consists of two diodes connected in parallel. It can be used in two different ways. If used one way, a varistor limits voltage excursions, and thus can protect the receiver from excessive signal voltages. For this purpose, the varistor V_3 is connected across the receiver and serves as a *peak clipper.*

If used another way, the resistance of a varistor varies inversely with the direct current flowing through it. The smaller the current, the higher the resistance. The direct current flowing down the telephone line varies with the length of the line. The longer the line, the less direct current is available to power the telephone instrument. If this effect were not compensated, the efficiency of the telephone transmitter would vary greatly with line length. The varistor V_1 is used to compensate for this effect, and it reduces the efficiency of the transmitter on short loops.

The balance network for the anti-sidetone circuit is formed by C_2, C_3, R_3, and V_2.

The telephone connects to the line from two terminals called the *tip* and the *ring.* This terminology comes from the early days of telephony when operators used a cord to connect one party's line to another party's line. The cord terminated in a plug, and the talking path was formed between the tip of the plug and a narrow ring around the plug. The terminology has remained.

TECHNOLOGICAL TRENDS IN TELEPHONE INSTRUMENTS

Telephone instruments continue to evolve in terms of reducing costs and of offering new features to users. This section describes the technology behind some of these trends.

Figure 2.24 illustrated the major elements in a telephone instrument. Also included are the housing of the instrument, the internal printed circuit board and components, and various cords and jacks (see Figure 2.26). Furthermore, modern features, such as a stored repertory of frequently dialed numbers, require some

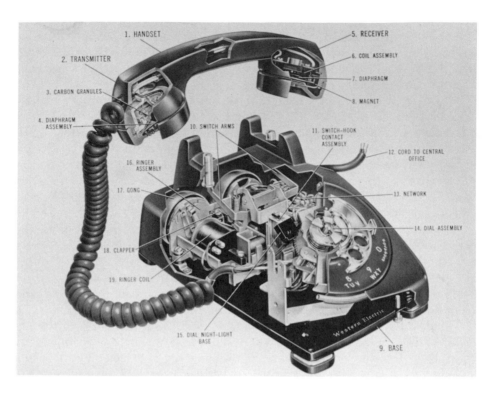

Figure 2.26 Cutaway drawing of an older Western Electric telephone instrument. (Courtesy of AT&T Bell Labs).

memory and processing in the form of a small microprocessor or computer. A visual display is sometimes used to display telephone numbers and to make the enhanced features easier for the user to control.

Edison's carbon transmitter is still in use in many telephones. The speech signal from the carbon transmitter can have a rough granular quality, but the rejection of background noise is quite good. Newer telephones use miniature self-polarizing capacitive microphones, called *electrets*. The speech quality is much improved, and noise-reduction circuitry is used to reduce the deleterious effects of background noise. The receiver in most telephones is still based on the original electromagnetic principles first used by Watson. Some newer telephones are starting to use piezoelectric receivers that offer advantages in weight, cost, and efficiency.

The switch-hook is usually a simple mechanical switch. Electronic switches and miniature relays are sometimes used in newer telephones. The ringer and dc voltages encountered by the switch are fairly large and could damage many conventional integrated circuits. High-voltage transistors can be used to perform the switch-hook function, and newer technologies allow them to be dielectrically isolated from other elements on the same integrated circuit, a technique called *hybrid integrated circuitry* (HIC).

The hammer-and-bell ringer invented by Watson is still used in most telephones. Newer telephones are using piezoelectric transducers and small loudspeakers. Electronic circuitry activated by the ringer signal creates the tones necessary for these transducers. The newer ringers are sometimes called *alerters*. They offer such advantages as small size, good efficiency, and nice sound characteristics. The circuitry required to generate the ringing signal is usually combined with the anti-sidetone network and the microprocessor on the same integrated circuit.

The old-fashioned rotary dial is fading away and is being replaced by push-buttons on a keypad. The keypad usually controls circuitry to generate either tones for touch-tone dialing or dial pulses. Mechanical keypads have been replaced by membrane and rubber-dome technology.

The telephone instrument should be easy to use and should have a pleasing physical appearance. The human factors of telephone instruments continue to be important. Telephones should be "user friendly" in all respects.

INSTRUMENT VARIETY

Telephone instruments come in all shapes and sizes with great variety. The simplest are black with a rotary dial and do nothing other than provide basic service. Complex instruments are cordless electronic telephones with built-in memory for a large repertory of telephone numbers.

Cordless telephones consist of a base station connected to the telephone line and a cordless handset that uses radio transmission to and from the base station. Cordless telephones are great when on the patio sipping a martini by the pool. The elimination of the cord can be a great convenience while cooking in the kitchen too. However, since radio transmission is used, anyone on the same frequency can listen to the conversation, thus making privacy an issue. Simple scrambling techniques are used on some newer cordless telephones to offer some protection against casual interlopers. The ringer is usually located in the base station because if the ringer were in the handset it might ring while the user was about to make a call and could damage the user's hearing.

Some telephones offer *hands-free* operation. The speakerphone accomplishes this through the use of a separate unit containing both the microphone and the loudspeaker. The microphone also picks up room noise and reverberation, which gives the speech that "rain barrel" quality. Acoustic feedback can occur between the loudspeaker and the microphone, but the switching of electric loss into the feedback circuit is frequently used to prevent feedback. Acoustic feedback and reverberation might someday be eliminated through the use of adaptive digital filters.

The simplest way to obtain hands-free operation is the use of a lightweight headset with the microphone located on a small boom. Such headsets are similar to those used by telephone operators and order takers. There is still a cord, however, so physical movement is restricted. A number of writers use such headsets while conducting interviews so that their fingers are free to type the interview into the word processor while the interview is being conducted.

Various "smart" features are available with many telephones. Last number redialing is a useful feature; the telephone remembers the number last dialed so that redialing a busy number is made much easier since only one key needs to be pushed. A repertory of frequently dialed numbers can be stored in the memory of some telephones and recalled simply by pushing a single key.

The cheap one-piece telephone was popular a few years ago. However, it quickly fell apart. Most consumers realize that the telephone instrument is an essential telecommunication device that must work day-to-day and particularly in time of emergencies. A quality, dependable telephone is clearly essential.

Station apparatus includes not only telephone instruments but anything that is connected to the telephone line. Facsimile machines for graphical communication, modems for computer communication, and telephone answering devices for recorded message taking are some other types of station apparatus.

STATION APPARATUS—AN ASSESSMENT

Some telephones resemble computers because of all their buttons and sophisticated features and displays. I remember a repertory telephone that had a cathode-ray tube display and could store thousands of telephone numbers. It was so complex that I doubt I could have used it to make a simple call.

The telephone instrument is basically a very simple device. New technology can be used to offer a great variety of sophisticated features, or it can be used to offer a simple, easy to use, quality product at a low price. All the most sophisticated features are useless if they are so complex that the average consumer cannot understand how to use them. I think of all the video cassette recorders that many consumers do not know how to program since the VCRs are too complex to fathom. Clearly, the marketing challenge is to develop in tomorrow's telephone instruments meaningful features that can be comprehended.

Chapter 3
TRANSMISSION

TRANSMISSION MEDIA

The aspect of telephony that deals with the various media and technologies for conveying signals—whether speech, data, or video signals—from one place to another is called *transmission.*

A variety of different transmission media are used to transmit and convey signals. Sometimes these media carry only a single signal, while other times they carry many signals combined together, or multiplexed, through either *frequency-division multiplexing* (FDM) or *time-division multiplexing* (TDM). The various media are as follows:

- open copper wire,
- paired wire (commonly called twisted pair),
- coaxial cable,
- microwave radio (terrestrial and satellite paths), and
- optical fiber.

These media vary greatly in the number of individual telephone signals, or speech circuits, that they can carry.

OPEN WIRE AND PAIRED CABLE

Open wire consists of uninsulated, bare copper wires that are strung on poles. The two wires that form the pair needed for the transmission of electric signals are physically separated on glass insulators by a distance of about one foot (1/3 meter) to prevent short circuits during high winds. Open wire is thick and therefore has a low loss, typically about 0.03 dB per mile. It was used during the early days of telephony until physical congestion became a serious problem. Open wire is still found, though very rarely, in rural areas.

A *twisted pair* is a pair of individually insulated copper wires twisted together with a full twist about every 2 to 6 inches. The insulation is usually plastic, but wood pulp has been used in the past. The diameter of the copper wires varies from 0.016 inch (26 gauge) to 0.036 inch (19 gauge). Many twisted pairs are combined into a single *cable,* usually sheathed with plastic as shown in Figure 3.1, although older cables were sheathed with lead. Anywhere from 6 to 2700 twisted pairs are combined together into a single paired cable. The gauge of the wire varies with the number of twisted pairs in the cable; finer wire is usually used in larger capacity cables.

Figure 3.1 A total of 1400 pairs of copper wire are carried in a single plastic-sheathed cable of about 3 inches in diameter. (Photo courtesy of AT&T Bell Labs.)

Paired cable can be strung on poles, buried underground, or installed in a conduit. A conduit consists either of long blocks of concrete with holes or plastic pipe through which the cable passes. Conduit is buried underground and offers the advantage that cable can be easily replaced, without digging up city streets, simply by pulling out the old cable and pulling through the new cable.

The thinner wire used in paired cable has a higher loss than open wire. The heaviest gauge wire (19 gauge) has a loss of about 1.1 dB per mile at 1000 Hz. The higher the loss, the smaller the signal that finally emerges after traveling through the wire. Popular paired cables contain 2700 pairs of 26-gauge wire, 1800 pairs of 24-gauge wire, and 110 pairs of 22-gauge wire.

A problem with many wire pairs all running parallel to each other with close spacing in a cable is that the electric signal on one pair can leak to another pair. This effect is called *crosstalk.*

Paired cable is used primarily for the local loop, and also between local exchange central offices. Baseband transmission is used on most local loops. In cases of severe congestion, *subscriber loop carrier* (SLC) systems are available, using either frequency-division multiplexing via amplitude modulation (AM) or time-division multiplexing via digital techniques.

Analog frequency-division multiplexing and digital time-division multiplexing are used on paired cables in exchange trunk transmission when congestion necessitates an enhancement of baseband transmission. These multiplexing schemes are used on short-haul (15 to 200 miles) trunks. The analog systems are called *N-carrier,* and typically multiplex together 24 voice circuits. The digital systems are called *T-carrier,* and multiplex together 24 voice circuits. Analog systems are very rarely used today, and most interoffice trunks utilize the digital T-carrier technology. T-carrier is described later in this chapter.

LOADING COILS

A serious problem during the early days of telephony was how to receive a strong signal over long lines. These were the days before the dawn of the electronic age and the amplifier. The early solution was the use of resonance to boost the signal along the line by the deliberate introduction of series inductance in the form of coils of wire called *loading coils.*

Loading coils were invented independently in 1899 both by Michael I. Pupin of Columbia University and by AT&T employee George A. Campbell. The final patent was awarded to Dr. Pupin based on a disclosure only two weeks earlier than Campbell's. The theoretical analysis on which the inventions were based was performed by Oliver Heaviside in England in the late 1800s. The first loading coils were installed experimentally in 1899 and were quickly adopted for use on most long lines. Even today, loading coils are used on long local loops, particularly in rural areas.

To understand why loading coils work, it is necessary to know that an electric transmission line can be modeled as a series of infinitesimally small elements with each element consisting of a series inductance L, a series resistance R, a shunt (or parallel) capacitance C, and a shunt resistance S (see Figure 3.2). These various

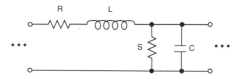

Figure 3.2 A transmission line is modeled as a series of very small sections, each section consisting of a series resistance and inductance and a parallel resistance and capacitance. These infinitesimal elements are usually expressed on a per unit length basis, for example, ohms per mile.

quantities are expressed in ohms, henrys, and farads on a per unit length basis, for example ohms per mile.

At frequencies encountered in voice telephone call, the attenuation A (or loss) of such a line is given approximately by the following equation:

$$A \approx \frac{R}{2} \sqrt{\frac{C}{L}} + \frac{1}{2S} \sqrt{\frac{L}{C}} \tag{3.1}$$

The first term represents the effects of the series losses, and the second term represents the effects of the shunt losses. Usually, the series losses predominate. Thus, the introduction of additional series inductance will decrease the predominant first term, resulting in a decrease in the overall attenuation of the line. Clearly, if too much series inductance is added, then the second term could become predominant, thereby negating the desired effect of reducing the overall attenuation. The required series inductance is added as discrete inductors placed in series every 6000 feet along the line. (Refer to Figures 3.3 and 3.4.)

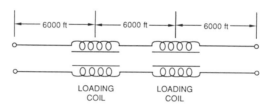

Figure 3.3 The series placement of loading coils reduces attenuation within the voice band. The loading coils are used on long loops and are spaced about every 6000 feet.

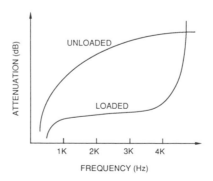

Figure 3.4 The effect of loading is to reduce attenuation in the voice band up to about 3500 Hz. However, the attenuation increases rapidly above that frequency. The use of loading coils sharply reduces the higher frequencies outside the voice band that can be transmitted along the line.

An inductor is a coil of wire, and since these series inductors load the line to reduce attenuation, they are considered to be loading coils. Their inductance is about 88 mH. The inductance is chosen to ensure a passband from 300 to about 3300 Hz. The type of loading most popularly used is specified as *H-88*. The *H* signifies a 6-kilofoot spacing, and the *88* signifies an inductance of 88 millihenrys.

Although a loading coil reduces attenuation in the voice band, attenuation outside this band is greatly increased. In essence, bandwidth is traded for reduced attenuation. The transition, or cut-off, frequency f_c is given approximately by the following equation:

$$f_c \approx \frac{1}{\pi \sqrt{LDC}} \tag{3.2}$$

where L is the inductance of the loading coil, D is the spacing in miles between the loading coils along the line, and C is the shunt capacitance per mile.

MULTIPLEXING

An early challenge to telecommunication inventors was that of discovering how to transmit more than one signal over a single wire circuit. The installation of wire circuits was very costly, and quite clearly great economies would result if more than one signal could be transmitted over each wire circuit. Among others, Thomas Alva Edison was attracted to solving this problem for telegraph signals; so too was Alexander Graham Bell in his investigations of the harmonic telegraph that ultimately led him to the invention of the telephone.

An early means through which telephone signals could share wire circuits was called the *phantom circuit*. The additional signal was carried in a balanced way over existing wire circuit pairs through the use of transformers at each end of the circuit, as shown in Figure 3.5. The technique could be extended to more wire circuits so that more phantom signals could be carried.

Today, advanced techniques are used so that a single transmission medium can carry thousands of combined signals. The sharing of a transmission medium by many signals is called *multiplexing,* as illustrated in Figure 3.6. Two approaches to multiplexing are analog, or frequency-division multiplexing, and digital, or time-division multiplexing. The actual equipment that performs the multiplexing is called a *channel bank.* Analog or A-type channel banks perform analog multiplexing; digital or D-type channel banks perform digital multiplexing.

Analog Multiplexing

In analog multiplexing, a number of voice circuits are combined, with each voice circuit given its own unique space in the frequency spectrum. Each telephone

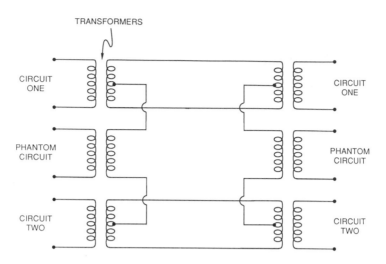

Figure 3.5 An additional phantom circuit is formed along two physical wire circuits through the use of transformers. The two wires of each physical circuit carry the phantom signal in a balanced way that has no net effect on signals carried over the physical circuits.

Figure 3.6 A multiplex unit combines many voice (and possibly data) signals to share a transmission medium or channel. The multiplexing can be accomplished through either frequency-division or time-division multiplex units.

speech signal occupies a frequency spectrum from about 200 to 3400 Hz; such a signal in its most basic form is called a *baseband signal.* Each baseband speech signal needs to be shifted, or translated, in frequency to a new unique band of frequencies. This frequency shifting is accomplished by amplitude modulation using *single-sideband, suppressed-carrier modulation.* The actual multiplexing is accomplished as a multileveled process in which a small number of signals are multiplexed together to form a group, and then a number of groups are multiplexed together, and so forth.

A baseband telephone speech signal contains frequencies in the range from 200 to 3400 Hz. A single voice circuit occupies a channel. Twelve channels are mutliplexed together, each channel being 4000 Hz wide, to create a *group.* A group covers the frequency range from 60 to 108 kHz. Five groups are multiplexed together to create a *supergroup* occupying the frequency range from 312 to 552 kHz.

The process can be continued (Figure 3.7). Ten supergroups multiplexed together give a *mastergroup* occupying the frequency range from 564 to 3084 kHz and containing a total of 600 voice channels. Six mastergroups multiplexed together give a *jumbogroup* occupying the frequency range from 564 to 17,548 kHz and containing 3600 channels. Three jumbogroups multiplexed together give a jumbogroup multiplex containing 10,800 channels.

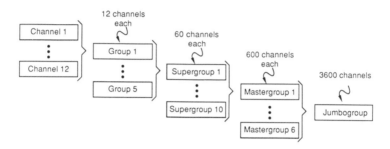

Figure 3.7 An analog multiplex unit, called an A-type channel bank, combines voice signals starting with a basic unit of 12 voice circuits, called a group. Frequency shifting through single-side-band, suppressed-carrier amplitude modulation is used.

Analog multiplexing and continual improvements in the technology enabled costly transmission systems to be shared to carry thousands of telephone signals, thereby making long-distance telephone service affordable to us all. But analog multiplexing suffered from noise, distortion, and other impairments and was costly to maintain. The age of analog multiplexing is now over; digital multiplexing has proved superior in nearly all respects and is now quite widespread. In the late 1980s, AT&T replaced nearly all the analog multiplexing in use on its long-distance network with digital multiplexing systems; MCI followed in the early 1990s.

Digital Multiplexing

Time-division multiplexing is used to combine a number of digital signals. A sample of each digital signal is given its own unique time interval or time slot, and a large number of time slots are then transmitted repetitively in this fashion. The digital signals can be either digital data generated by a computer or telephone speech that has been converted from an analog signal into a digital format.

The first step in converting an analog speech signal into a digital format is to *filter* the signal to prevent false, or *alias,* frequencies from appearing later upon reconstruction of the analog signal. The telephone speech signal therefore is low-pass filtered to 4 kHz. The next step is to *sample* the band-limited signal at the Nyquist rate of 8000 samples per second. The corresponding sampling interval is 1/8000 second, or 125 μs.

The sampled signal is then *quantized* into either 256 or 128 levels, depending on the number of bits that will next be used to encode each quantized sample. Eight bits are used with 256 levels; seven bits are used with 128 levels. Usually, eight bits per sample are used. The bit rate corresponding to eight bits per sample is 64,000 bits per second. A 64-kbit/second digital version of a single voice signal is called a DS0 signal. Since the amplitudes of the signal are encoded as binary combinations of on-off pulses, digital is also known as *pulse code modulation.*

The sampled signal is usually *compressed* in amplitude and is then quantized using a linear quantizer having equal-sized quantization steps. This allows more resolution or steps for smaller amplitudes of the signal because speech signals rarely have very large amplitudes. The signal will be later *expanded* upon reconstruction to obtain the original dynamic range. The process of compression followed by expansion is called *companding.* The compression characteristic is logarithmic with more compression as the amplitude of the signal increases. The μ-law is used in North America and Japan, while the rest of the world uses the A-law. Conversion is thus required on international calls between these two parts of the world.

Twenty-four voice channels time-division multiplexed together give a DS1 signal, requiring a bit or data rate of 1.544 million bits per second (Mbps). A group of 24 voice channels digitally multiplexed together is sometimes called a *digital group,* or a *digroup* for short. As shown in Figure 3.8, four DS1 signals time multiplexed together give a DS2 signal, containing 96 voice channels and requiring a data rate of 6.312 Mbps. Seven DS2 signals time mutliplexed together give a DS3 signal, containing 672 channels and requiring 44.736 Mbps. Six DS3 signals time multiplexed together give a DS4 signal, containing 4032 channels and requiring a

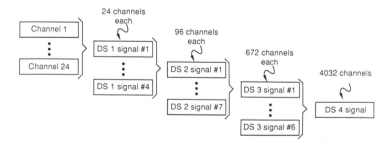

Figure 3.8 A D-type channel bank combines voice signals through time-division multiplexing. The basic multiplexing unit consists of 24 voice channels, sometimes called a *digroup.* A single voice channel is digitized at 64,000 bits per second and is sometimes referred to as a DS0 signal. A digroup, or 24 multiplexed voice signals, is called a DS1 signal.

data rate of 274.176 Mbps. The extra bits in the higher capacity digital signals are used for timing and synchronization information to assist in the separation and demultiplexing of the individual channels.

MODULATION AND MULTIPLEXING

Multiplexing is the combination of a number of signals to share a transmission medium. Analog modulation is the shifting of signals in frequency through either the amplitude modulation or the frequency modulation of a pure-tone carrier wave. Digital modulation is the sampling and encoding of an analog signal as a series of binary pulses.

The world of transmission has gone digital but that does not mean that analog modulation is dead. Most transmission media carry such analog waves as light or radio signals. This means that although time-division multiplexing is used to multiplex together a number of digital signals, the multiplex signal must then be transmitted over the transmission medium using analog modulation of a carrier wave. For example, thousands of voice channels are time-division multiplexed together for transmission over optical fiber. But the digital multiplex signal ultimately is used to turn on and off the analog wave of light that is transmitted along the fiber—a form of analog amplitude modulation. As another example, the digital multiplex signal transmitted over microwave radio ultimately must modulate an analog radio-frequency carrier wave that shifts the digital multiplex signal into the frequency band of the assigned microwave radio channel.

Most transmission media ultimately carry analog waves. But these analog waves might represent digital multiplex signals that contain thousands of speech signals that have been time-division multiplexed together. The analog wave is varied in amplitude, frequency, or phase in synchrony with the multiplex signal. The one exception is baseband transmission of a digital signal in which the actual multiplex signal is carried directly as an electric signal over the medium, usually copper wire in the form of twisted pair or coaxial cable.

COAXIAL CABLE

Analog multiplexing on *coaxial cable* has been used in the old Bell System since 1946 for long-distance telephone transmission. A number of one-way voice circuits are frequency multiplexed together using single-sideband, suppressed-carrier amplitude modulation on a single coaxial pipe, or simply a coaxial. Two such coaxials make a two-way pair, with each coaxial carrying transmission in one direction. As the signal travels along the coaxial, it becomes weaker and weaker and must be amplified before it becomes too weak. Amplification is a one-way affair, and thus a coaxial can only carry signals in one direction. A number of coaxials are placed together to form a coaxial cable for use in a transmission system. The photo of Figure 3.9 shows a coaxial cable used in the latest-generation L5 carrier system.

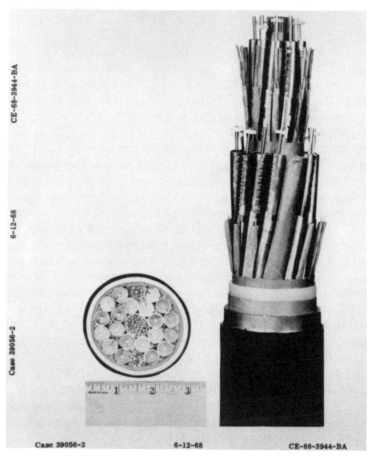

Figure 3.9 Cutaway view of the coaxial cable used in the L5 system. Ten pairs of coaxial pipes formed the cable. (Photo courtesy of AT&T Bell Labs.)

The multiplexing system used with coaxial cable is called *L-carrier.* The various generations of the technology are indicated by a number suffixed after the L. A key factor in the L-carrier systems is the distance between the amplifiers, called *repeaters,* at which the signal is amplified and retransmitted down the next section of the cable. As repeater and mutliplexing technology improved, the distance between the repeaters could be decreased at a reasonable cost, and the overall capacity of the system could be increased.

The diameter of each coaxial is 3/8 inch. Because the capacity of the system is large, one coaxial pair in the cable is kept as a spare to be used in case of trouble with one of the pairs in service.

Figure 3.10 shows the progression over time of the L-carrier system. The most recent system is L5E, which was first placed in service in 1978. The L5E system used integrated circuit technology with repeaters spaced every mile along the route. As with all L-carrier systems, the cable is buried underground. The channel capacity of a single coaxial in the L5E system is 13,200 channels (22 mastergroups). There are 11 coaxial pairs, 10 of which are in actual service. Thus, the overall capacity of the L5E system is 132,000 two-way voice circuits. Coaxial cable is being replaced by optical fiber.

	L1	L3	L4	L5	L5E
SERVICE DATE	1946	1953	1967	1974	1978
TECHNOLOGY	vacuum tube	vacuum tube	transistor	integrated circuit	integrated circuit
REPEATER SPACING (miles)	8	4	2	1	1
CAPACITY PER COAX					
in channels	600	1860	3600	10,800	13,200
in groups	mastergroup	3 master & 1 supergroup	jumbogroup	jumbogroup multiplex	22 mastergroups
COAX PAIRS	4	6	10	11	11
WORKING PAIRS	3	5	9	10	10
ROUTE CAPACITY (two-way voice circuits)	1800	9300	32,400	108,000	132,000

Figure 3.10 The generations of the L-carrier coaxial cable system.

MICROWAVE TERRESTRIAL RADIO

Basic Principles

Radio transmission is used to carry telephone conversations across continents and oceans. Different frequency bands have been allocated for use by the common carriers for the transmission of telephone signals. A large number of voice circuits are multiplexed together in these radio systems.

The microwave radio bands are in the gigahertz (10^9) range of frequencies and are used in cross-country terrestrial and satellite routes. The extremely high frequency waves are transmitted between radio antennas. The radio waves are conducted to and from the antennas through metal pipes called *waveguides*. Two microwave radio bands in current use are from 3.7 to 4.2 GHz, called the 4-Ghz band, and from 5.925 to 6.425 GHz, called the 6-GHz band. The width of each of these bands is 500 MHz, and they are both used for microwave terrestrial radio. Each band is subdivided into a number of channels. The width of each channel in the 4-GHz band is 20 MHz, and the width of each channel in the 6-GHz band is 30 MHz.

Microwave radio beams follow a line-of-sight path. This means that radio towers for the antennas need to be located about every 26 miles along the route of the system to take into account the curvature of the earth, as shown in Figure 3.11. An antenna on each tower receives the radio signal, which is shifted to a new channel, amplified, and retransmitted to the next tower (see Figure 3.12). The towers perform as repeater stations. For the 4- and 6-GHz bands, rain does not usually have a significant effect on the propagation of the radio wave. However, rain does have a significant effect on the higher frequency 11-GHz (10.7 to 11.7 GHz) and 18-GHz (17.7 to 19.7 GHz) bands, which are also used for microwave transmission.

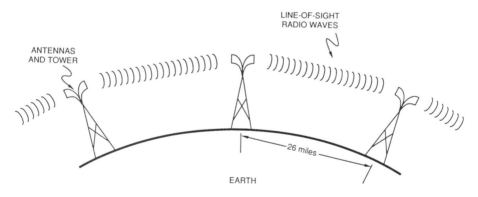

Figure 3.11 In terrestrial microwave radio systems, radio signals travel from tower to tower, each tower located on a line-of-sight path 26 miles from the next.

Microwave terrestrial radio used to be the backbone of the long-distance network and is still a fairly important technology. Analog frequency-division multiplexing was used extensively with microwave radio systems in the past. The increasing use of optical fiber has supplanted microwave radio to a considerable extent, but existing microwave routes have been converted to digital time-division multiplexing, thereby giving renewed life to an old technology. We now review the history of microwave radio; it is an exciting story of how advances in technology continually led to ever increasing channel capacity and lower rates to the consumer.

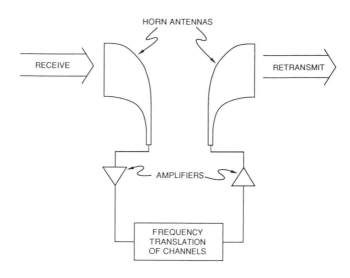

Figure 3.12 At each microwave tower, the received signal is changed in frequency to a different radio channel, amplified, and retransmitted to the next tower.

TD Radio System

The first cross-country microwave radio system in the AT&T long-distance network became available in 1950. It used the 4-GHz band and was called TD-2. The 500-MHz bandwidth was divided into 25 channels, each 20 MHz wide (see Figure 3.13). Because of mutual interference, adjacent channels could not be used. Two channels were needed to create a two-way circuit with one channel in each direction, and thus the maximum number of two-way channels was only six. Five of these six two-way channels were used for service, and one was reserved as a spare.

	FREQUENCY RANGE	BANDWIDTH	CHANNEL WIDTH	NUMBER OF CHANNELS
4 GHz BAND	3.7 to 4.2 GHz	500 MHz	20 MHz	25
6 GHz BAND	5.925 to 6.425 GHz	500 MHz	30 MHz	16

Figure 3.13 The two microwave bands most often used for long-distance transmission. The 4-to-6-GHz band is the most popular because it is least affected by propagation and noise problems.

Noise immunity was a problem, and hence frequency modulation with its greater immunity to noise and interference was used to modulate the radio-frequency carrier for transmission. However, frequency modulation trades noise immunity for bandwidth, and hence only 480 voice circuits could be transmitted over each radio channel. The overall capacity of the TD-2 system was five two-way radio channels with each channel carrying 480 voice circuits for a total system capacity of 2400 two-way voice circuits. In 1953, the capacity of each radio channel was increased to 600 voice circuits for a total system capacity of 3000 two-way voice circuits. This total capacity was soon doubled through an advance in technology.

Radio waves can be *polarized* so that they propagate as either vertical or horizontal waves. By polarizing one radio channel vertically and the adjacent radio channel horizontally, interference between adjacent channels was eliminated, and all 24 radio channels were utilized. This dual polarization was accomplished with the invention of a special horn antenna, shown in the photograph of Figure 3.14.

Figure 3.14 Photograph of microwave relay tower used in AT&T long-distance network. (Photo courtesy of AT&T Bell Labs.)

The net effect was to double the total system capacity in 1959 to 6000 two-way voice circuits. The truly exciting aspect of this development was that the same physical tower and route were used simply by adding new antennas and other electronics.

In 1968, a solid-state transistorized microwave radio system operating in the 4-GHz band was introduced. This system, called TD-3, had 1200 voice circuits per radio channel and a total system capacity of 12,000 two-way voice circuits. The final version of TD technology using analog frequency-division multiplexing of the voice circuits had 1800 voice circuits per radio channel. Of the total of 12 two-way radio channels, 11 were in service and 1 was for protection, giving a total capacity of 19,800 two-way voice circuits.

TH Radio System

The horn antenna first used in the late 1950s was capable of handling radio signals in both the 4-GHz band and the 6-GHz band. Hence, a microwave radio system operating in the 6-GHz band could "piggyback" on the existing 4-GHz TD system. The 6-GHz system was called TH radio. The first TH system became available in 1961 and utilized vacuum-tube technology. As shown in Figure 3.15, by utilizing

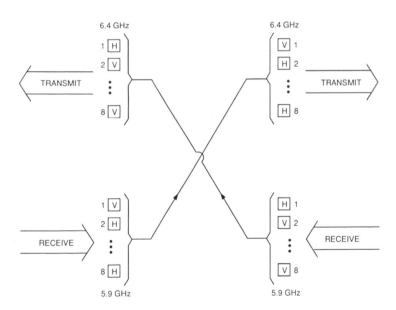

Figure 3.15 The use of vertical and horizontal polarization of the radio waves doubled the capacity of a microwave radio system by reducing interference between adjacent channels. In the TH system, the signal received in the 5.9-GHz portion of the band is retransmitted to the next tower in the 6.4-GHz portion of the band.

dual polarization, all 16 radio channels could be used. This gave a total of 8 two-way radio channels, 6 of which were in service. The TH system had a capacity of 1800 voice circuits per radio channel and a total system capacity of 10,800 two-way voice circuits. The TH system was used with an already existing TD system for a total combined capacity on the route of 16,800 two-way voice circuits. In 1979, improved amplifiers and other technological advances enabled the capacity of the TH system to be increased to 2400 voice circuits per radio channel. But this improvement was short lived.

AR6A Radio System

In 1981, AT&T introduced the AR6A system as a replacement for the TH system. The major innovation in the AR6A system was the use of single-sideband, suppressed-carrier, microwave transmission of the radio signal with its inherent increase in bandwidth efficiency when compared to the previously used frequency-modulation transmission. This meant that, with AR6A, 6000 voice circuits could be placed in each radio channel. The total of seven active two-way radio channels available in the 6-GHz band, with one two-way channel for protection, gave a total system capacity of 42,000 two-way voice circuits. Because AR6A was used in addition to the existing TD system, the total maximum route capacity equaled the 42,000 circuits of AR6A plus the 19,800 circuits of TD for a total of 61,800 two-way voice circuits.

However, the days of frequency-division multiplexing were soon to be over, and this advance would soon be supplanted by digital time-division multiplexing.

Short-Haul Systems

The number of retransmissions, or *hops,* of the microwave radio signal distinguish *short-haul* microwave systems from *long-haul* systems. Short-haul systems typically have from 1 to 10 hops, and long-haul systems have from 10 to 50 hops. Microwave systems that cross the continent are long-haul systems.

Microwave radio is also used for short-haul communication utilizing the 6-GHz and 11-GHz bands. The 6-GHz system is called TM, and the 11-GHz system is called TL. Each radio channel in these systems carries 1200 voice circuits utilizing analog frequency-division multiplexing. To prevent interference between the received and retransmitted signals, 12 channels in the 11-GHz band are used for a given hop, and the remaining 12 channels are used in adjacent hops. The net result is that 6 two-way channels are available for a total capacity of 7200 two-way voice circuits for the 11-GHz TL system.

Digital Microwave Radio

In the early 1980s, AT&T introduced digital radio systems that used time-division multiplexing and operated in the microwave radio bands for use in short- to medium-length routes. The DR6-30 system was first used in 1981 and operates in the 6-GHz band with channels that have a bandwidth of 30 MHz. A total of 1344 two-way digital voice circuits are transmitted in each of seven two-way radio channels. This corresponds to two DS3 signals per radio channel at a rate of about 90 Mbps. The total system capacity using all seven radio channels is 9408 two-way digital voice circuits. The DR6-30 system is used for hops from 15 to 30 miles.

At about the same time, AT&T also introduced the DR11-40 system operating in the 11-GHz band with 40-MHz channels. The ten two-way radio channels carry a total of 13,444 two-way digital voice circuits. The higher frequency 11-GHz band is more susceptible to fading caused by rain, and hence the DR11-40 system is restricted to hops of 15 miles or less.

The time-division multiplexed signal is transmitted over the air using a technique called *quadrature amplitude modulation* (QAM). The digital signal is transmitted as a continuous series of short bursts of a radio-frequency wave. Each short burst, called a *baud,* can vary both in maximum amplitude and in phase. A total of 16 combinations of amplitude and phase are possible for each baud, as shown in Figure 3.16, and thus each baud can encode four bits. This specific type of quadrature amplitude modulation is called 16-QAM. The challenge with QAM is the ability to distinguish reliably one combination of amplitude and phase from another in order to know precisely what information was transmitted. Noise, fading, and interference all conspire to make this task difficult. Nevertheless, AT&T improved the technology and developed 64-QAM, which encodes and transmits six bits per baud with an error rate of less than 1 in 10 billion bits.

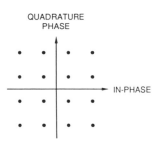

Figure 3.16 Each short burst of the radio wave can have 1 of 16 possible combinations of amplitude and phase. The specific combination determines the four bits that were transmitted for that short burst. This type of digital transmission is called *quadrature amplitude modulation* (QAM).

The DR6-30-135 system operating in the 6-GHz band and utilizing 64-QAM was introduced by AT&T in 1984. It transmits three DS3 digital signals in each radio channel at a rate of about 135 Mbps. The total system capacity of the DR6-30-135 system is 14,112 two-way digital voice circuits, and the system replaces frquency-division multiplexing systems in the long-distance network. A DR11-40-135 system operating in the 11-GHz band with a total system capacity of 20,160 two-way digital voice circuits is also available.

Digital radio systems operating in the 4-GHz band will soon be available along with further increases in system capacity. The DR4-20-90 system operating in the 4-GHz band is capable of carrying two 45-Mbps signals per radio channel. With 11 active two-way radio channels, the DR4-20-90 system is able to carry 14,784 two-way digital voice circuits.

The older DR18 digital microwave radio system operating in the 18-GHz band has been used on very short hops of from one to five miles. This is because of attenuation of the radio signal caused by moisture in the air. Each channel in the microwave band operates at a 274.176-Mbps data rate, corresponding to 4032 digital voice circuits. A total of eight two-way radio channels is used with an additional two-way channel as protection for a total system capacity of 28,224 two-way digital voice circuits. *Differential phase-shift keying* (DPSK) with *dibit* coding is used to modulate the radio carrier in each channel.

COMMUNICATION SATELLITES

Basic Principles

Terrestrial microwave radio cannot go everywhere. It is very costly to construct a series of towers to reach very remote areas with few telephone users. Also, terrestrial microwave radio cannot cross oceans because radio waves travel along a line-of-sight path and floating microwave towers are not feasible. The solution is a single microwave "tower" placed high enough in the sky so that great distances can be covered in a single hop up to and down from the tower. Since neither sky-hooks nor towers hundreds of miles high are feasible, some practical form of implementation is needed.

The practical solution is the use of an artificial earth satellite for communication purposes. This solution was first proposed by the renowned science fiction author Arthur C. Clarke in 1945 (in *Wireless World*). Scientists at Bell Labs and elsewhere become intrigued with Clarke's idea. By the end of the 1950s, a team of scientists at Bell Labs under the technical direction of John R. Pierce had constructed some prototype systems. The first communication satellite, *Telstar I*, was launched on July 10, 1962.

The early system proposed by Bell Labs would have consisted of a large number of satellites at the relatively low altitude of 3000 miles. At this altitude, satellites orbit the earth very quickly, and hence a series of satellites would need to be tracked continuously across the sky as they passed overhead. Such a complex system had serious technical and economic problems. The more practical solution was a single satellite orbiting the equator at a height such that the orbit time of the satellite was exactly the same 24 hours as the daily rotation of the earth. In this way, the satellite would appear stationary with respect to the surface of the earth if the satellite were orbiting the earth's equator.

The physical principles that explain satellites are fairly simple. (Refer to Figure 3.17.) Imagine a cannon located at the top of a tower and aimed parallel to the ground. If a shot is fired from the cannon, the shot will travel a distance from the cannon at the same time as it falls back to the ground. If more charge is added to the cannon, the shot will travel farther, but will ultimately fall back to the ground. If a sufficient charge is placed in the cannon, the shot will have enough velocity so that the curvature of the earth's surface will be falling away at the same rate as the shot is being pulled toward the ground by gravity. The shot will then continuously fall around the earth. It will have become an artificial satellite of the earth.

The height of the orbit of a satellite and its rate of rotation about the earth are related. The higher the orbit, the slower the rate of rotation about the earth. There is a height at which the rate of rotation of the satellite is exactly the same 24 hours that it takes for the earth to complete one full revolution about its axis. If the

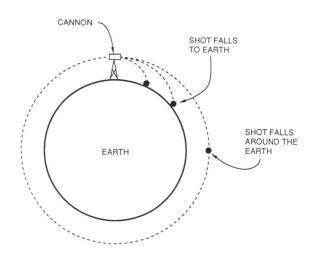

Figure 3.17 A communication satellite continuously falls back to earth but at the same rate as curvature of the earth falls away from it. The satellite thus orbits the earth.

orbit of the satellite is exactly above the earth's equator, is circular, and is in the same direction as the earth's rotation, the satellite will appear stationary with respect to the earth's surface. This type of orbit is called a *geosynchronous orbit,* and the height of such a circular orbit is 22,300 miles above the earth's equator, or equivalently 26,300 miles from the earth's center. (See Figure 3.18.)

As stated earlier, the use of geosynchronous satellites for communication was first suggested in 1945 by Arthur C. Clarke. The first geosynchronous satellite for intercontinental communication, *Syncom II,* was launched in July 1963. This satellite spun about its axis to achieve stability in orbit and was conceived in 1959 by Harold A. Rosen and designed by his team of engineers at Hughes Laboratories.

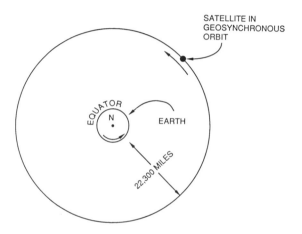

Figure 3.18 A communication satellite in a circular orbit 22,300 miles above the earth's surface takes 24 hours to complete one full orbit. If located above the earth's equator, it thus appears stationary with respect to the earth turning at the same rate beneath it. This is called a *geosynchronous orbit.*

Electronic circuitry on the satellite receives the signals transmitted to it from the earth station. These signals are very weak and must be amplified by low-noise amplifiers (LNAs) carried on the satellite. The signals are then shifted to a new frequency band and are retransmitted back to earth. The frequency bands are divided into a number of radio channels, and a separate radio signal is sent in each channel, as with terrestrial microwave transmission. The amplification, frequency translation, and retransmission of the signal for each channel are performed by circuitry called a *transponder* on the satellite. Each radio channel has its own transponder, and hence the satellite must have a number of transponders to cover the whole frequency band assigned to it. Modern communication satellites operating in the most popular bands typically carry 24 transponders.

Frequency Bands

High-frequency microwave transmission is used to send signals up to communication satellites and then back to earth. The microwave band of frequencies used to transmit from the earth to the satellite is called the *uplink*, and the band of frequencies used to transmit from the satellite back to the earth is called the *downlink*, as shown in Figure 3.19. The frequency bands must be different to eliminate interference between the uplink and the downlink.

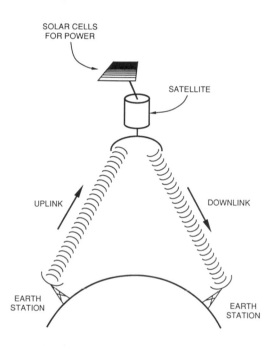

Figure 3.19 The signal transmitted from an earth station to a communication satellite is called the *uplink*, and the signal transmitted back to earth is called the *downlink*. Usually, a single antenna on the satellite with different feeds is used for both the uplink and the downlink.

The earliest frequency bands used for satellite transmission were the same 4-GHz and 6-GHz bands used for terrestrial microwave transmission. The 4-GHz band (3.7 to 4.2 GHz) is used for the downlink, and the 6-GHz band (5.925 to 6.425 GHz) is used for the uplink. The 4-GHz and 6-GHz bands used together for satellite communication are called the C band, or sometimes the 6/4-GHz band. Some newer communication satellites use the Ku band operating at 12 GHz (10.95 to 11.2 GHz plus 11.45 to 11.7 GHz) for the downlink and 14 GHz (14.0 to 14.5

GHz) for the uplink. International agreements have authorized the use of a third band, the Ka band, at 17 and 30 GHz.

Because only a very few number of bands are available for satellite communication, the same bands must be shared by all communication satellites. Thus, to prevent interference, it is essential that the radio signal sent to one satellite not be received by an adjacent satellite operating in the same band, and similarly the earth station aimed at one satellite must not receive the signal from an adjacent satellite. This means that the transmitted radio beam and the aperture of the receiving antenna must be very narrow. However, these widths depend on the frequency band used for transmission and are a factor in determining the permissible spacing between the satellites in geosynchronous orbit (see Figure 3.20). The permissible spacing in the C band has been 4 to 5 degrees, which translates into a distance of about 2000 miles between adjacent satellites. The higher frequency Ku and Ka bands allow the beam of the radio signal to be more narrow than the C band. With the Ku band, the distance between satellites can be decreased to 2 degrees, thereby allowing more satellites to serve the same geographic area. However, the higher frequency bands are more subject to absorption of the radio signals by rain.

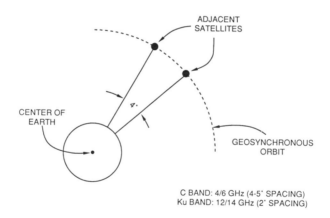

C BAND: 4/6 GHz (4-5° SPACING)
Ku BAND: 12/14 GHz (2° SPACING)

Figure 3.20 Communication satellites are spaced apart in geosynchronous orbit. The angle subtended with the center of the earth is about 5 degrees for the 4/6-GHz C-band. The use of the higher frequency Ku band allows the spacing to be reduced to 2 degrees, but propagation problems are much worse in this band.

The typical spectrum width of the radio channel served by one transponder is 36 MHz, allowing for some guardband space. Because reception of a weak signal in the presence of noise is a prime consideration in satellite communication, frequency modulation with its inherent noise immunity is used for modulating the microwave radio signals. A single transponder can be used for one color television signal, 1200 voice circuits, or digital data at a rate of 50 Mbps.

The total width of each half of the C band is 500 MHz. Horizontal and vertical polarization of the radio signal is used so that 24 channels are available for use. A communication satellite allocates these channels to create 12 two-way transponder pairs.

Delay

Geosynchronous communication satellites are at a considerable distance above the surface of the earth, and hence a fair amount of time is required for the radio signal to reach the satellite from the earth and to be retransmitted back. Radio waves travel at nearly the speed of light, which is 186,000 miles per second. A radio wave will therefore take about 120 milliseconds (ms) to travel the 22,300 miles that the satellite is above the earth. Since the earth station is not directly below the satellite, a little extra time needs to be added for this extra distance. A net time of 135 ms thus is reasonable for the radio wave to travel between the earth station and the satellite.

The signal received by the satellite is retransmitted back to earth. An additional 135 ms is required for the retransmitted radio signal to travel that distance. Thus, there is a total time of about 270 ms for a signal to be sent by satellite from one location on the surface of the earth to another, assuming that a single satellite hop is used.

For a telephone conversation, one person speaks and the other person responds. Using a satellite, it would take 270 ms for the speech to travel from one person to another and an additional 270 ms for the repsonse to travel back. The total round-trip time before the first speaker hears the beginning of the reponse thus is 540 ms, or about one half second. This amount of time is perceived as a delay in response, is quite noticeable to most people, and is quite bothersome to many. If one were using a computer terminal to a distant computer via a satellite link, this round-trip delay would add appreciably to the response time. This delay would be intolerable if one computer were communicating interactively with another at high speeds.

In some communication situations, one satellite hop might be connected back-to-back to another satellite hop thereby doubling the round-trip delay. The delay in such multihop satellite links can bring interpersonal communication to a virtual halt, unless each person learns to wait patiently for the other person's response and never to interrupt the other person. A possible solution to multihop links would be to beam signals directly from one satellite to another, thereby eliminating the need for intermediate retransmission to the earth. The use of lasers would be particularly applicable to this type of intersatellite communication because there is no optical interference in the vacuum of space.

Access

The size and shape of the satellite's radio beam on the surface of the earth are called the *footprint* of the satellite. Clearly, for a satellite that is broadcasting signals to as large an audience as possible, such as in direct broadcast satellite transmission of television signals to homes, the footprint should be large. If the satellite is sending signals to only a specific earth station, then the footprint should be small. In the first case, the radio beam would need to be broad; in the second case, the radio beam would need to be narrowly focused. The antenna design and available transmitting power on the satellite are important factors in determining footprints.

The satellite's antenna can be designed so that it can send narrowly focused beams to a number of specific, geographically dispersed locations. Furthermore, the narrowly focused beam can be switched from one location to another so that information can be transmitted to one location and then to another. If broad beams are used to reach multiple locations, then everyone receives the same information and security may become a problem. Narrowly focused beams, switched or unswitched, avoid this problem.

A number of separate earth stations may wish to beam information to the same satellite. One way this can be accomplished is for each multiple user to be assigned a specific range of frequencies within a transponder channel for each transmission. This type of access is called *frequency-division multiple access* (FDMA). Another way to accomplish multiple access is to assign specific times, or time slots, to each user during which time the information must be transmitted. This type of access is called *time-division multiple access* (TDMA) and is particularly well suited to the bursty nature of data communication.

In addition to the actual communication signals relayed by a satellite, *telemetry* signals are transmitted to and from the satellite. These signals report on the satellite's temperature, battery life, orbital position, and other parameters. If the position of the satellite needs to be corrected because of drift, signals from the earth will activate small thrusters on the satellite to reposition it.

Satellites are powered by solar energy from light rays from the sun. This light energy is converted into electric energy by panels of solar cells attached to the satellite. The efficiency of this solar conversion of energy determines the total electrical power available to the satellite for transmission of signals back to the earth among other purposes. For a portion of each orbit, the satellite will be shielded from the sun by the earth. During this time, batteries previously charged from the solar panels are used to power the satellite.

Capacity

The communication satellite *Comstar I* operating in the 6/4-GHz C band was launched in 1976. It carried 24 transponders, and each transponder channel had a capacity of 1200 voice circuits. Utilizing horizontal and vertical polarization of the radio waves, the total capacity of the satellite was 14,400 two-way voice circuits. *Comstar IV* was launched in 1980. Its total capacity was increased to 21,600 two-way voice circuits through improvements in technology that enabled each transponder channel to carry 1800 voice circuits.

Frequency modulation was originally used for the radio signals transmitted to both *Comstar I* and *Comstar IV*. In 1982, the equipment at the earth stations was changed to use single-sideband amplitude modulation (SSB-AM) of the radio wave. SSB-AM is considerably more spectrum efficient than frequency modulation, and hence the number of voice circuits handled by each transponder channel was increased to 7800 for a total capacity of 93,600 two-way voice circuits. This significant increase in capacity was equivalent to that achieved through the similar use of SSB-AM in the AR6A terrestrial microwave system.

The overall capacity of communication satellites could be further increased through the use of SSB-AM that may allow as many as 12,000 voice circuits per transponder channel. However, the trend is toward digital time-division multiplexing of voice circuits with a reduction of overall capacity for a given transmission bandwidth. This is because the digital speech signal of 64,000 bps has a bandwidth of about 32,000 Hz even though the bandwidth of the analog speech signal is only 4000 Hz.

The capacity of communication satellites for carrying data is quite impressive. As an example, *Comstar IV* has 24 transponder channels, each capable of carrying 670 Mbps. The overall capacity of the satellite thus is 1.4 Gbps. This equates to transmitting all the text in 1.1 million 200-page books in one hour. Although this may seem impressive, and indeed is, optical fiber has far greater capacities.

Satellites—An Assessment

The delay problem inherent in satellite transmission is a serious disadvantage for interpersonal telephone conversation. One way of minimizing the problem is to use a satellite circuit for transmission in one direction and a terrestrial circuit for the return transmission. Satellites simply are not that suitable for voice transmission unless the transmission is to a very distant place that is otherwise unreachable.

Communication satellites do offer considerable bandwidth at a transmission cost that is independent of distance. They are quite suitable for the one-way broadcast transmission of television signals between various television network locations and from the network to affiliated stations. Nearly all network and cable television broadcasters rely heavily on satellites for the distribution of program material.

A single satellite could broadcast television directly to our homes. A *direct broadcast satellite* (DBS) utilizing the 12-GHz band (12.2 to 12.7 GHz) would transmit a signal that could be received by a small two-foot antenna. Although the technology for DBS is in hand and some other countries have already launched direct broadcast satellites, there are many nontechnical issues that cloud the future of this technology for the United States. The local affiliates of the major networks would be completely bypassed. Regional broadcasting would die. The "big brother" image of a single broadcasting antenna in the sky would frighten many politicians. Most importantly, one wonders whether there is much more of a market for more sources of televison programs and video given the widespread penetration of cable television and VCRs.

Satellite technology continues to advance. Rocketry and space technology have advanced so that large payloads can be placed in orbit quite economically. Competition in rocket launching from private industry, as advocated by some people, could lower launch costs even more. Satellites need to carry fuel for orbital repositioning. Ion beams could accomplish this repositioning using electric power obtained from solar panels. The amplification of the signal broadcast back to earth by the satellite is performed by traveling wave tubes (TWTs). The lifetime and reliability of these tubes will continue to improve, thereby extending the useful life of the satellite. The present 4-degree spacing will decrease to 2 degrees so that more satellites can share the same geosynchronous orbit.

T1 DIGITAL CARRIER

Basic Principles

Digital carrier systems utilize time-division mulitplexing to combine a number of digital signals. The digital microwave radio systems previously described are examples of digital carrier systems. Other transmission media can be used to carry time-division multiplexed digital signals. The digital signals usually are speech signals encoded at 64,000 bps but can also be data transmitted between computers and computer terminals. Voice telephony generates considerably more traffic than all other sources, and hence most of the traffic carried in digital form is from voice circuits.

The earliest use of digital carrier, or pulse code modulation, was the T1 system, which first saw service in 1962. The T1 digital carrier system is used on short-haul interoffice trunks less than 50 miles long. Conventional 16- to 26-gauge copper wire pairs are used, with a wire pair needed for each direction. The digital information is transmitted directly as a baseband electrical signal over the wire. With distance, the on-off pulses that form the baseband digital signal become smeared together. Before it is no longer possible to distinguish one pulse from another, the pulses are detected and are regenerated again for transmission over another length of the wire pair. This process is performed by electronic devices, called *regenerative repeaters,* that are located every mile along the wire pair.

Twenty-four digital voice circuits are time-division multiplexed together in the T1 system. Each digital signal consists of a series of 8-bit samples. The 24 digital signals multiplexed together thus become a string of 8 × 24 or 192 bits, called a *frame,* as shown in Figure 3.21. The demultiplexing equipment at each end must be very precise in decoding the fast string of bits so that the 8 bits for each circuit are assigned and decoded correctly. To allow recovery from any mishap, an extra bit is added to each string of 192 bits. This 193rd bit alternates from on to off each frame. This regular pattern of bit alternation can be used by the demultiplexing equipment to identify the 193rd bit and thus regain synchronization. The 193rd bit is therefore called the *synchronization bit.*

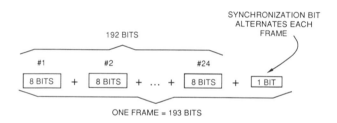

Figure 3.21 In the T1 digital carrier system, 24 digital circuits are combined. Each digital circuit contains 64,000 bps corresponding to the analog speech signal that has been sampled 8000 times per second with 8 bits being used to encode each sample. All the groups of 8 bits for each of the 24 circuits are combined together sequentially and are transmitted as a series of 192 bits, called a *frame.* A 193rd bit is added for synchronization purposes. Frame after frame is transmitted at a rate of 1.544 Mbps.

All 193 bits that form a frame must be transmitted in 1/8000 second, or else the bits from the incoming digital circuits will start to accumulate. Thus, the overall bit rate is 193 bits per 1/8000 second, or 1.544 Mbps. Each individual bit must be sent in 1/1,544,000 second, or 0.65 microsecond (μs).

One of the eight bits in each digital voice channel is used every sixth frame for signaling and supervisory purposes, for example, to specify that a particular channel is idle or in use.

Bipolar Pulses

The digital information in the T1 system is transmitted in the form of *bipolar pulses*. A binary 0 is encoded as zero volts. A binary 1 is encoded alternately as a positive pulse or as a negative pulse, as shown in Figure 3.22. These pulses have a width equal to one-half the bit time of 0.65 μs. The alternating polarity of the pulses gives an average voltage of zero to the digital signal and also makes it simple to detect a single-bit error.

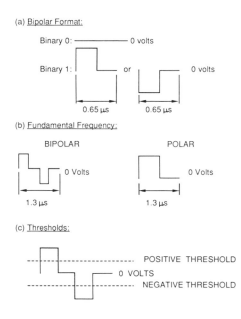

(a) Bipolar Format:

(b) Fundamental Frequency:

(c) Thresholds:

Figure 3.22 (a) The bipolar format used in the T1 system is a return-to-zero scheme in which a binary 0 is encoded as zero volts, and a binary 1 is encoded as a return-to-zero pulse that alternates in polarity from pulse to pulse. (b) Even though the bipolar pulses are half the width of polar pulses, the maximum fundamental frequency of a string of pulses is the same. (c) The bipolar format requires two threshold decisions at the receiver.

If conventional polar pulses with a width of 0.65 μs were used, the maximum fundamental frequency of a pulse train with alternating zeros and ones would be about 772 kHz. The use of pulses with half that width would normally double the fundamental frequency and hence the bandwidth needed to pass that signal. However, the bipolar format has its maximum fundamental frequency for a string of binary ones, and that frequency is also about 772 kHz. Thus, the bandwidth requirements are not increased by the use of bipolar pulses. However, because there are three voltage levels with the bipolar format, a decision based on two thresholds must be made at the receiver to determine the binary information. This is more complicated than with the simpler polar format, where a decision based on only one threshold must be made because there are only two possible voltage levels.

Expanded Systems

As we saw previously with digital microwave radio, a number of DS1 digital signals can be multiplexed together and transmitted over any medium. The bandwidth of the medium will determine the overall data rate and capacity that the medium can support.

Conventional wire pairs (twisted pair) are used as the transmission medium for the T1 system. Special 22-gauge wire pairs with low electrical capacitance are used to create a digital carrier system with more circuit capacity than the conventional T1 system. This is because wire pairs with low electrical capacitance have a greater bandwidth than conventional wires and thus can pass a higher data rate. This system is called T2 digital carrier.

Four DS1 signals are time-division multiplexed to create the DS2 signal that is used in the T2 system. T2 was first introduced into the Bell System in 1972 for use on intercity circuits up to 500 miles in length. Regenerative repeaters are spaced about every 3 miles. Special low-capacitance 22-gauge wire is used. A total of 96 digital voice circuits is carried on each wire pair in a T2 system. Usually, 24 wire pairs are used in a cable, with one wire pair reserved as protection. The total capacity of a system is 2208 two-way digital voice circuits, using a separate cable in each direction. Each digital T2 line operates at a bit rate of 6.312 Mbps. The bipolar format is used.

In the early 1980s, AT&T introduced an improved T-carrier system, called T1G, operating on standard twisted-pair cables. The T1G system carries 96 digital voice circuits in each direction over two twisted pairs. Rather than the bipolar pulses used in the older T1 system, the newer T1G system uses four-level pulses, called a *quaternary format.* Each one of the four possible levels represents one combination of two bits. For example, a voltage level of $+3$ volts represents the binary digits 11; $+1$ volt represents 01; -1 volt represents 00; and -3 volts represents 10.

Digital carrier is also used on coaxial cable. Coaxial cable with its much higher bandwidth than conventional wire provides for even higher data rates. The T4M digital carrier system uses coaxial cable and operates at 274.176 Mbps, corresponding to 4032 digital voice circuits. A T4M system uses the same coaxial cable consisting of 11 pairs of coaxials used in the L5 system. One pair of coaxials is reserved for protection. The total capacity of the T4M system is 40,320 two-way digital voice circuits. The T4M system typically has been used on routes of 500 miles or less in length. A polar format is used with the T4M to provide increased noise immunity compared to the bipolar format used in the T1 and T2 systems.

OPTICAL FIBER

Basic Principles

Imagine a very long mirror that has been rolled up to create a long tube with the mirrored surface on the inside of the tube. Alternatively, imagine a long pipe with an inside surface that is mirrored. If a source of light shines into one end of the tube or pipe, the light waves will be repeatedly reflected off the surface of the mirror and guided along the length of the pipe until they emerge at the other end. In effect, the mirrored tube is a light, or optical, waveguide. The use of mirrored pipes to carry light is an old idea: the American William Wheeler was granted a patent in 1881 on the use of mirrored pipes for guiding light.

There is a better way of creating a mirror to reflect light at shallow angles. An optical boundary is formed when two materials having different indices of refraction are joined. A light ray passing across this boundary is bent when it emerges in the second material, as illustrated in Figure 3.23. The angle at which the light ray is bent is proportional to the relative indices of refraction of the two materials—a principle known as Snell's law of refraction. If the light ray impinges on the boundary at a very shallow angle, called the *critical angle,* it will be reflected at the boundary as if the boundary were a perfect mirror.

Figure 3.23 (a) A light ray passing across an optical boundary is bent. (b) If the angle of incidence is less than some critical angle, the light ray is reflected at the boundary.

Consider a long, thin piece of glass made in the form of a fiber. The interior *core* of the fiber is surrounded by a *cladding* of glass with a lower index of refraction than the core. If a light wave enters the core at a shallow angle, it will be repeatedly reflected at the boundary of the core and the cladding until it emerges from the end. This is the basic principle of the optical fiber. It is a medium for transmitting light waves. It too is an old idea: the use of thin glass fibers for guiding light was described in 1887 by the British physicist Charles Vernon Boys.

The idea of using light as a transmission medium for speech signals was conceived in 1880 by Alexander Graham Bell with the invention of his photophone. It transmitted voice waves over the air on light waves. The synthesis of Bell's ideas with the light-carrying capabilities of thin glass fibers did not occur then. It was not until quite recently that optical fibers have come into widespread use for communication. Digital multiplexing was not known a century ago nor was the required electronic circuitry available. But optical fiber is indeed the transmission medium of choice today for telecommunication circuits across continents, under oceans, and between central offices. Someday, optical fiber may also be the choice for local circuits all the way from the central office to homes and offices.

The exterior of an optical fiber is coated with a *jacket* (see Figure 3.24) to prevent stray light from entering the fiber. The actual glass from which the optical fiber is manufactured is silica with an extremely high purity. Typical losses are about 0.2 dB/km, and fibers with even lower losses are under development as the technology continues to progress. As a comparison, optical glass has a loss in the order of about 1000 dB/km.

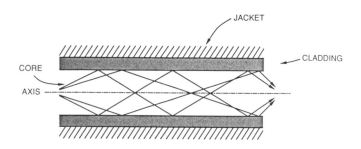

Figure 3.24 Light waves entering an optical fiber at less than the critical angle are continuously reflected through the glass core of the fiber.

An optical fiber communication system consists of a light source, the fiber itself, and a light detector at the distant end, as shown in Figure 3.25. The light pulses are detected and regenerated along the length of the system at regular intervals determined by the degree of smearing, noise, and losses in the fiber. The regeneration of pulses along the length of the system is called repeating and is performed by electronic devices called regenerative repeaters.

Light is the carrier wave for signals multiplexed together and transmitted over optical fiber. Today it is difficult to modulate the frequency or amplitude of light in a highly linear fashion. But what is very simple to do is to turn the light on and off. Optical fibers thus are particularly well suited for carrying digital signals that

Figure 3.25 The light from a source is guided through the optical fiber to a detector at the distant end. This forms a basic fiber communication system.

have been time-division multiplexed together. Light has an extremely high frequency and can be turned on and off at an extremely fast rate, and hence a very large amount of information can be carried over a single optical fiber. This large capacity combined with their freedom from noise and interference, low cost, low loss, and small size are some of the reasons why optical fibers have truly revolutionized transmission technology in such an exciting way.

Types of Optical Fiber

Light waves traveling through an optical fiber take different paths depending on their specific reflections. All these paths have different lengths, with the shortest path being right down the axis of the fiber. These different paths mean that the light will arrive at different times depending on the length of the paths. Hence, a sharp pulse of light will arrive at slightly different times and will thus be smeared in shape (see Figure 3.26). The amount of smearing, or pulse spreading, determines the rate at which data can be sent over the fiber. Some types of optical fiber are better than others in terms of smearing.

Three types of optical fiber are defined by optical grading of the fiber itself and the *modes* that the light travels during transmission. The concept of mode is concerned with the number of possible light paths through the fiber. Multiple-

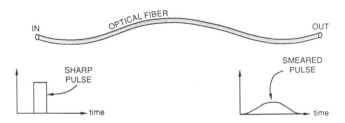

Figure 3.26 Because the light travels through the fiber on many paths of different lengths, a sharp pulse is smeared in time when it leaves the fiber. Losses in the fiber will attenuate the amplitude of the pulse. These effects limit the maximum capacity of the fiber.

mode, or *multimode,* fiber has many possible paths along which the light travels and thus is subject to much smearing.

The simplest type of multimode fiber consists of an inner core with a uniform index of refraction surrounded by an outer jacket with a lower index of refraction. This type of optical fiber, illustrated in Figure 3.27, is called *multimode stepped-index* fiber. The inner core is about 50 μm in diameter, or roughly the diameter of a human hair. If a sharp pulse of light is sent along the fiber, it emerges quite smeared because of the many different paths it travels. This smearing limits the maximum pulse rate to about two million pulses per second and the practical distance between repeaters to about five kilometers.

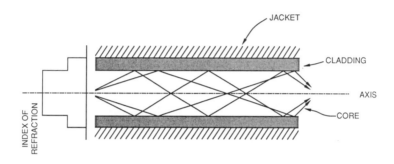

Figure 3.27 Refractive index and some representative light paths in a multimode stepped-index optical fiber.

Light travels in a mutlimode manner through the second type of optical fiber, but the refractive index decreases gradually away from the axis of the core. The result of this graded index is that light rays are gradually turned back toward the axis of the core as they travel through the fiber. This type of fiber, illustrated in Figure 3.28, is called *multimode graded-index* fiber. Another way of visualizing the workings of multimode graded-index fiber is that the light rays are continuously refocused as if the fiber were a continuous succession of small lenses. The diameter of the core of multimode graded-index fiber is also about 50 μm. Because of the continuous refocusing of the light rays, smearing of pulses is much less. The practical pulse rate for multimode graded-index fiber is about 100 million pulses per second.

The third type of optical fiber is called *single-mode* fiber (see Figure 3.29). The diameter of the core of single-mode fiber is very small, typically about 5 μm. Only parallel rays of light can pass through it by traveling along its axis, and multiple paths with their inherent smearing are avoided. More will be said later in this section about the capacities of optical fiber.

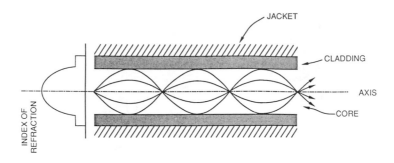

Figure 3.28 Refractive index and some representative light paths in a multimode graded-index optical fiber. The light is continually refocused as it travels along the fiber.

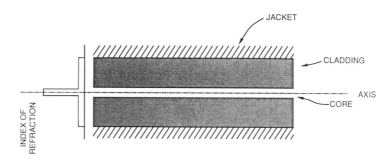

Figure 3.29 Single-mode optical fiber has a core that is so thin that only parallel rays of light travel along it. In this way, all the rays of light arrive at the end at the same time, and smearing is nonexistent.

Source and Detector

If the speed of an optical fiber system is less than 100 Mbps, then a simple *light-emitting diode* (LED) can be used as the light source in an optical fiber communication system. LEDs are economical and very reliable.

If the speed is more than 100 Mbps, then a laser must be used as the light source. The light output of a laser is a collimated beam of single-frequency coherent light. The collimated beam of parallel light rays makes it easy to couple the beam to single-mode fiber. Lasers used to be large and costly with shorter lifetimes than LEDs. However, all that changed with the advent of solid-state miniaturized lasers, described later in this chapter.

A light-sensitive semiconductor diode is usually used as the light detector. Semiconducting material is photosensitive, and its resistance varies with the intensity and frequency of the light stimulating it. The resistance decreases as the illuminating light increases in intensity. A semiconductor diode is used as a *photodiode* by allowing light to illuminate the semiconducting material at or near the depletion region by the junction. The diode is reverse biased, and the leakage current varies according to the light absorbed at the depletion region. Photosensitive diodes are sometimes made from a sandwich of positive-doped, intrinsic (or natural), and negative-doped semiconducting materials. Such diodes are called *PIN diodes.*

If the reverse bias is increased to a high voltage, the electrons freed by the light will collide with other atoms and will have sufficient energy to create hole-electron pairs which, in turn, will collide with other atoms creating more hole-electron pairs, and so on, in an avalanching fashion. Such diodes are thus called *avalanche diodes.* The effects of the multiplication of the current carriers give them an internal gain and an increased sensitivity. Avalanche diodes are also used as photodiode detectors in optical fiber communication systems.

Lasers

A single-mode optical fiber communication system uses a laser as its light source, and hence we should understand the principles of operation of lasers. A laser generates a beam of light with all the light rays parallel to each other. This property of laser light is called *collimated.* Furthermore, the light waves are all in phase with each other, a property called *coherent.* Last, the light usually is a single frequency, a property called *monochromatic.* Lasers generate a collimated beam of coherent monochromatic light.

The principles of operation of lasers involve quantum mechanics. (Refer to Figure 3.30.) If an atom is excited by an external source of energy, the atom will absorb some of the energy and some electrons will move into higher energy bands. The atom is not stable in this excited state and will return to its normal unexcited state rather quickly. When this happens, the excited higher energy level atoms fall back to their normal levels. In doing so, they release precise quanta of energy in the form of light at a precise frequency, or a set of precise frequencies, depending on the particular type of atom. This release of energy in the form of light is called *spontaneous emission.* It is the way fluorescent lamps work.

The emisison of light energy can also be stimulated by further exciting an already excited electron with additional light energy at precisely the correct frequency. This form of light emission is called *stimulated emission* and forms the basis of operation of a laser.

A laser consists of a chamber, called a *resonator* (or resonant chamber), containing a material whose atoms have the property that they can be excited to pro-

(a)

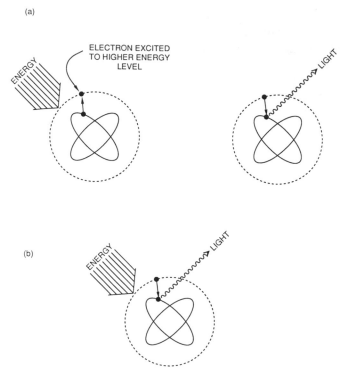

(b)

Figure 3.30 (a) Spontaneous emission of light occurs when an electron is excited to a higher energy level by an external source of energy and then returns spontaneously from this unstable level to its normal, stable level. (b) An electron already excited to a higher state can be further excited by external energy to return to its stable state, and the stimulated emission of light occurs during this stimulated change of energy level.

duce laser light (see Figure 3.31). The material is said to be able to *lase* and could be either a solid or a gas. A mirrored surface is placed at each of the two parallel ends of the chamber. An external source of energy is then applied to the material in the chamber. This causes the atoms to be excited, and some of them emit energy in the form of the spontaneous emission of light. The light energy thus produced from spontaneous emission then stimulates adjacent already excited atoms to emit light through stimulated emission. This light then travels to other adjacent atoms causing the further stimulated emission of light in a cascade fashion. Quickly, a wave of light energy travels through the material in the chamber until the wave encounters one of the mirrors.

When the mirror is encountered, the light wave is reflected and travels back through the material in the chamber, thereby stimulating the emission of more light

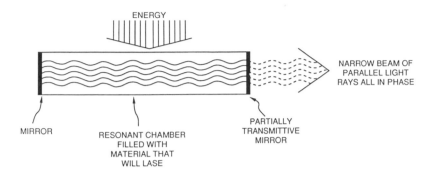

Figure 3.31 In a laser, external energy produces the stimulated emission of light. The light is reflected from mirrors at each end of a resonant chamber, thereby creating an optical standing wave that stimulates the emission of more light. A portion of the light passes through a partially transmittive mirror at one end of the chamber. The laser light is a monochromatic, coherent, collimated beam of light.

energy. The mirror at the other end of the chamber is encountered, the light is reflected again, and the release of more light is stimulated in the chamber. The continuously reflecting wave of light quickly reaches a fairly high intensity, and the process continues as long as the external source of energy is applied. The distance between the mirrors is chosen to create an optical resonant chamber, and therefore an optical standing wave is created in the chamber. One of the mirrored surfaces is partially transmissive, and thus some of the light can pass through the mirror and emerge from the chamber. The process involves *l*ight *a*mplification by the *s*timulated *e*mission of *r*adiation, hence the term *laser*.

Because there is one wave traveling back and forth in the resonant chamber, the light wave is coherent in phase. Also, the stimulated emission produces light of a single, precise frequency. That frequency, or color, is dependent on the atomic properties of the material in the chamber. Some materials will lase at more than one frequency. The laser light is generated by many internal reflections within the chamber, and hence the light rays are all parallel to each other. Because all the light is concentrated in a single beam, the beam of light can be quite intense. Laser light can be so intense that it can be used as a surgical knife and can even burn holes in metal.

Since not all atoms will lase, the material in the chamber must be the correct kind to produce laser light. The material can be a gas, a solid, or even a liquid. Some gaseous materials used in lasers are helium-neon, carbon dioxide, and argon. Ruby is one solid material that has been used in lasers. The semiconductor material used in a diode will lase too. The small size and low power requirements of semiconductor diodes when used as lasers make them particularly well suited for consumer electronics and communication applications.

We are already familiar with the LED used for visual displays in clocks and other appliances. The type of light emitted by a LED is conventional light. However, if the diode is constructed of the proper semiconducting material with a junction in the form of a small resonant chamber and if the junction is forward biased beyond a threshold current, the diode will emit laser light from the junction, as shown in Figure 3.32. Laser diodes usually use gallium arsenide (GaAs) as the semiconducting material. The junction forms the optical resonant chamber. One end of the junction is made optically reflective by a metallic coating, and the other end is partially transmissive so that the laser light can escape. The other sides of the chamber are cut rough to inhibit lasing in the wrong direction. The threshold current is about 100 milliamperes (mA).

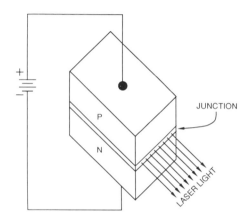

Figure 3.32 A semiconductor diode will emit laser light from its forward-biased junction. The junction is constructed in the form of a small optically resonant chamber.

Capacity

The increase over time in the capacity of commercially available optical fiber communication systems is astonishing. AT&T's first fiber system, FT3, was introduced in 1979 and used graded-index fiber at 45 Mbps capable of carrying 672 voice circuits (one DS3 digital channel). A total of 72 fiber pairs was placed together in a single cable, and 66 of the pairs were in service with the remainder as spares. The total route capacity of the system was 44,352 two-way digital voice circuits.

The *FT3C* system was introduced soon afterward with each fiber operating at 100 Mbps and carrying 1344 digital voice circuits. Using cable with 66 active pairs, the system had a maximum route capacity of 88,704 two-way digital voice circuits. Repeaters were located every four miles along the route.

In 1984, AT&T introduced the *FTX-180* system with each single-mode fiber operating at 180 Mbps and carrying 2688 digital voice circuits. A 400-Mbps system was introduced in 1986 with each single-mode fiber carrying 6048 digital voice circuits and repeaters spaced every 20 miles. Fiber systems operating at 2000 Mbps (2 Gbps) are standard.

The capacities of today's fiber systems are so mind-boggling that something other than Mbps needs to be used. I like to remember that 50 Mbps is equivalent to transmitting all the text in 8 paperback books in one second, or carrying 780 digital voice circuits, or carrying one color television signal. Using this method of equivalency, 2000 Mbps is equivalent to 320 paperback books per second, 31,250 digital voice circuits, or 40 television signals. This is most impressive but is nothing compared to the theoretical maximum capacity of single-mode fiber.

The theoretical maximum capacity of a single light channel in single-mode fiber is about 200 Gbps, that is 200 *billion* bits per second (200×10^9 bps). This capacity could carry 30,000 paperback books per second, 3 million digital voice circuits, or 4000 television signals. If the entire light spectrum operable for fiber (the 1.45- to 1.65-μm wavelength region) was used through color multiplexing, then the theoretical capacity increases to 50 Tbps, or 50,000 Gbps (50×10^{12} bps). This capacity could carry 7.5 million paperback books per second, 750 million digital voice circuits, or 1 million television signals. All these impressive capacities are theoretical maxima, and we are nowhere near such capacities, nor would we want to be. Clearly, it would make little sense to put all our signals in one fiber basket because, in case of trouble, the damage would simply be far too great.

Optical Fiber—An Assessment

One advantage of optical fiber is its low loss, which means that very few repeaters are needed along a fiber route. For example, in August 1983, Bell Labs scientists gave a laboratory demonstration of an unrepeatered 100-mile length of fiber operating at 420 Mbps. Clearly, an unrepeatered fiber strand under the Atlantic will someday be possible.

One problem with fiber was that the many vendors all had systems that did not interconnect. This meant that the optical signal had to be demultiplexed and then multiplexed again when two fiber systems were joined. The solution is an interface standard that enables the use of standardized rates operating synchronously so that the transmitter and receiver are locked together at a fixed rate of transmission. This standard is called SONET for *Synchronous Optical Network.* The basic SONET rate in the United States is 51.84 Mbps; in Europe, it is 34 Mbps. Multiples of these rates, such as 155.52 Mbps, have been standardized, too.

Optical fiber offers many advantages over other transmission media. The fiber itself is flexible and inexpensive. Because light is carried by the fiber, optical fiber is not prone to electrical and electromagnetic interference. Because no light escapes from the fiber nor can enter it from outside, there is no leakage or crosstalk between adjacent fibers. The bandwidth and capacity are extremely high, and losses are quite small.

The extremely small size of the fiber—one-tenth the diameter of a human hair for single-mode fiber—makes splicing very tricky, but techinques have been developed to speed this task. However, switching must be accomplished by converting the optical signal to an electrical signal and then using conventional techniques. Similarly, amplification and regenerative repeating are accomplished by converting the light signal to an electrical signal. In both cases, it would perhaps be more efficient if these tasks could be accomplished directly on the light signal without the need for conversion to an electrical signal. Research is under way to solve these problems.

Single-mode fiber clearly is the winner. By using a number of different light signals at different frequencies, the overall capacity of a single strand of fiber can be increased—a technique called *color multiplexing*. Semiconductor lasers are tiny and inexpensive and are the obvious choice as light sources in fiber systems.

TRANSOCEANIC UNDERSEA CABLE

Cable is laid under the sea for communication across oceans. Clearly, it is difficult and costly to make repairs of cable and repeaters located far under the ocean. Hence, reliability is the prime consideration in transoceanic undersea cable systems. This reliability is obtained by the use of conservative technology and by a trade-off against system capacity. Continual improvements in repeater design allowed closer spacing and more repeaters to be used, with a resulting increase in system capacity, as shown in Figure 3.33.

The first transatlantic coaxial cable system, TAT-1, was installed by AT&T and went into service in 1958. It consisted of two coaxials, called SB-type cable, one in each direction, and had a total system capacity of only 36 two-way voice circuits. A total of 51 electron-tube repeaters were used with this early system, and the underwater repeaters were spaced every 44 statute miles.

The next generation of submarine cable also utilized vacuum tubes, but carried 138 two-way voice circuits on a single SD-type coaxial. This system was first used in 1963 on the TAT-3 transatlantic system between the United Kingdom and New Jersey, with repeaters spaced every 23 statute miles. A single SF-type coaxial and transistorized repeaters were used in 1970 on the TAT-5 transatlantic system, with repeaters spaced every 12 statute miles. TAT-5 was designed to carry 845 two-way voice circuits.

The circuit capacity of these early, and later, submarine systems could be greatly increased through the use of a technique called *time-assignment speech*

TRANSATLANTIC SUBMARINE CABLE SYSTEMS

SYSTEM NAME	SERVICE DATE	CABLE		REPEATERS			TWO-WAY VOICE CIRCUIT CAPACITY	
		TYPE	TECHNOLOGY	QUANTITY	SPACING (STATUTE MILES)	TECHNOLOGY	SYSTEM DESIGN	MAXIMUM TASI
TAT-1	1958*	SB	TWO COAXIALS (0.62 IN)	51	44	VACUUM TUBES	36	72
TAT-2	1959*	SB	TWO COAXIALS (0.62 IN)	57	44	VACUUM TUBES	36	72
TAT-3	1963*	SD	ONE COAXIAL (1.0 IN)	182	23	VACUUM TUBES	138	276
TAT-4	1965*	SD	ONE COAXIAL (1.0 IN)	186	23	VACUUM TUBES	138	345
TAT-5	1970	SF	ONE COAXIAL (1.5 IN)	361	12	TRANSISTORS	845	2112
TAT-6	1976	SG	ONE COAXIAL (1.7 IN)	694	6	TRANSISTORS	4000	10,000
TAT-7	1983	SG	ONE COAXIAL (1.7 IN)	660	6	TRANSISTORS	4200	10,500
TAT-8	1988	SL	THREE FIBER PAIR	108	41	LASERS AND INTEGRATED CIRCUITS	8000 DIGITAL	40,000

*RETIRED FROM SERVICE

Figure 3.33 The history of transatlantic submarine cable systems installed by AT&T shows a steady increase in circuit capacity obtained primarily through improvements in repeater design allowing more repeaters at closer spacings. With optical fiber, however, the future trend will be toward fewer repeaters more distantly spaced. (Courtesy of AT&T International Cable Engineering and Construction.)

interpolation, or TASI. TASI uses the silent intervals in speech conversations to carry signals from other speech conversations. This requires rather complicated and costly processing at each end of the submarine cable system, but the increase in circuit capacity is well worth the additional cost. As you can well imagine, it is quite costly and difficult to install submarine cable systems under oceans. With TASI, the early TAT-1 system could carry 72 two-way voice circuits, and the later TAT-5 system carries 2112 two-way voice circuits.

The transatlantic optical cable system TAT-8 was completed by AT&T in 1988 and utilized SL-type cable consisting of three pairs of single-mode optical fiber in a single cable, with one pair for protection. Each fiber pair carries 4000 64-kbps two-way digital voice circuits. Digital processing that capitalizes on the silent intervals in speech conversations, digital TASI, is used to increase the maximum system capacity to 40,000 two-way voice circuits. The repeaters are spaced every 41 statute miles, and 108 repeaters are used along the total length of the cable. AT&T's newest

TAT-9 system, scheduled for service in 1991, will have repeaters spaced every 75 statute miles and will carry a maximum of 80,000 two-way voice circuits.

Undersea cable is laid down by cable ships, such as AT&T's *C. S. Long Lines,* shown in Figure 3.34. The cable is usually placed in trenches by trenching machines when installed along the continental shelf to protect the cable from trawling or other damage. However, for other parts of the route, the cable is simply laid out from the ship and allowed to settle on the floor of the ocean. Today's cable ships are kept quite busy laying cable under the Atlantic and the Pacific oceans and other seas. The large capacity and reliability of optical fiber make it the transmission medium of choice, and rates for international calls will continue to decrease as capacity continues to increase.

Figure 3.34 Much of the optical fiber of the TAT-8 system was installed under the Atlantic by AT&T's cable ship *C. S. Long Lines,* shown here at dock. (Photo courtesy of AT&T Bell Labs.)

SUMMARY OF TRANSMISSION CAPACITIES

The various transmission media that are currently available for long-distance telephone communication are:

- coaxial cable,
- terrestrical microwave radio,

- satellite microwave radio,
- optical fiber, and
- undersea cable.

The maximum route capacity of each of these media has grown over time at an increasing rate of growth, as shown in Figure 3.35.

This is truly an exciting story of the ability of technology to offer continuous improvements resulting in ever-increasing route capacities. For the consumer, this has created a true bargain in long-distance communication.

The newest medium—and clearly already the winner—is optical fiber. The capacity of fiber continues to grow, and the distance between repeaters continues

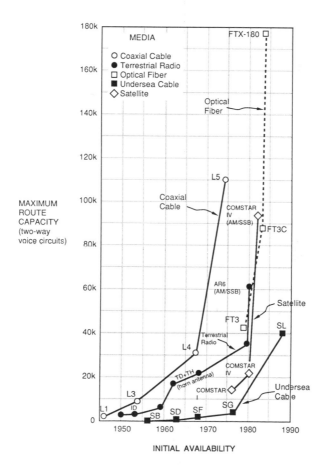

Figure 3.35 The maximum route capacities of the various transmission media used in the AT&T long-distance network have increased greatly over time.

to increase. One can only wonder whether fiber transmission will be so cheap that long-distance rates will someday be distance independent.

FOUR-WIRE–TO–TWO-WIRE CONVERSION

The local loop is a two-wire circuit. As shown in Figure 3.36, the induction coil in the anti-sidetone circuit in the telephone instrument separates the signal generated by the transmitter from the signal intended for the receiver. The two-wire circuit from the transmitter and the two-wire circuit to the receiver constitute a four-wire circuit. The induction coil thus converts a two-wire circuit into a four-wire circuit. Another name for the induction coil, or transformer, that performs such conversions is a *hybrid*.

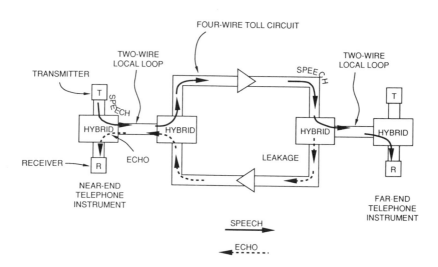

Figure 3.36 Hybrids match the two-wire local loop to the four-wire circuits used in toll transmission. The hybrid cannot perfectly match the impedance of every possible local loop, and hence some portion of the signal leaks through the hybrid and returns down the four-wire circuit in the opposite direction. Furthermore, electrical reflections occur along the local loop, and these reflected signals too are passed back down the four-wire circuit along with the signal that leaks through the hybrid. The returned signal is heard as an echo.

Long-distance and toll circuits are unidirectional, requiring separate paths for the signals traveling in each direction. This is because the amplifiers and repeaters used in long-distance and most interoffice trunk carrier systems are inherently unidirectional. These circuits thus are four-wire circuits, and they must be matched to the two-wire circuits used in the local loop. This matching is accomplished by hybrids. A hybrid is required at each end of a toll circuit. The hybrid at the end near the speaking party is called the *near-end hybrid;* the hybrid at the other end is called the *far-end hybrid.*

Each hybrid used in the toll network consists of two transformers with taps on the primary and the secondary. A far-end hybrid is shown in Figure 3.37. A balance circuit attempts to match the impedance of the two-wire local loop. Consider a signal coming into the hybrid from the four-wire circuit. This signal passes through the primary of the incoming transformer generating identical voltages at the two halves of the tapped secondary. One of these voltages generates a current that passes down the local loop and returns through one half of the tapped primary of the outgoing transformer. The other voltage generates a nearly identical current that flows through the balance circuit and returns through the other half of the tapped primary of the outgoing transformer. The reversed leads of the balance circuit mean that the currents flowing through each half of the primary of the outgoing transformer are out of phase with respect to each other. The voltages induced at the secondaries are also out of phase and cancel. Thus, no current flows from the outgoing transformer to the four-wire circuit.

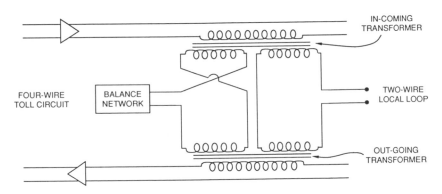

Figure 3.37 The hybrid is constructed from transformers. A signal entering the far-end hybrid from the four-wire circuit appears only at the two-wire circuit and does not return through the four-wire circuit. The hybrid function can also be performed by transistorized circuitry, but since the balance network can never match every local loop perfectly, a small amount of return signal still flows.

If hybrids were ideal, they would perform perfectly, and no signals would leak through them to return down the four-wire circuit. To perform perfectly, the two currents in the primary of the outgoing transformer must match perfectly. For this to occur, the matching circuit must match perfectly the impedance of the local loop. Because each local loop has a different impedance depending on its length and other factors, such perfect impedance matching is not possible. Thus, some current will leak across the hybrid, and some of the input signal will be returned back down the four-wire circuit toward the near end and will be heard by the speaking party. If the four-wire circuit is long, the returned signal will arrive with a delay and will be heard as a distinct echo by the speaking party.

If the impedances are not matched perfectly at various discontinuities along a transmission line, electrical reflection of a signal traveling down the line will occur. Discontinuities occur at electrical connections, and there are usually a number of them along a local loop. Thus, a number of small reflections occur for a signal traveling along a local loop, and these reflections travel back along the loop and appear as additional echo to the speaking party.

The returned signal can leak through the near-end hybrid and travel along the four-wire circuit to the far-end hybrid where it could leak again, and so forth. The net effect is to give the echo a reverberant quality, as if the person were speaking in a barrel. All this is most annoying, and for some amounts of delay, normal speech is impossible. But there is a solution to the problem of echo.

ECHO ELIMINATION

Echo can be a serious problem on toll circuits, and, therefore, it must be eliminated. One solution is to break or open the return path in the four-wire circuit so that the echo signal cannot return to the near end. This is accomplished by devices called *echo suppressors.*

As illustrated in Figure 3.38, an echo suppressor prevents the echo from returning down the four-wire circuit by either completely opening the return circuit or by inserting sufficient loss in the return circuit to reduce the echo signal to an inaudible level. This prevents echo caused by either leakage through the hybrid or reflections along the local loop. A control circuit senses when a signal is at the input to the hybrid and then activates the loss in the return path to prevent any echo. A *compare circuit* compares the input to the hybrid with its output to determine when the loss should be removed from the return path.

Early echo suppressors completely opened the return circuit. This meant that the other party's speech could never be heard during periods of double talking, and the circuit became half-duplex. Delay was deliberately designed into the dynamics of the control and compare circuits so that abrupt changes would not occur during the normal silent pauses in human speech. What all this meant was that unless the speaking party stopped talking entirely for enough time for the return circuit to be closed, the other party could not capture the return circuit.

Improved echo suppressors insert loss in the return circuit, which allows some of the other party's speech to be heard during double talking. But echo suppressors

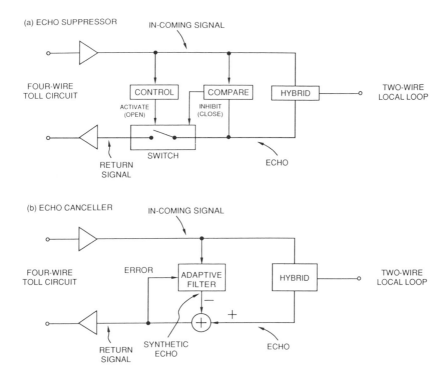

Figure 3.38 (a) The echo suppressor opens the return path and thus prevents echo. Unfortunately, it also prevents the speech from the other party from being heard during double talking. (b) The echo canceller uses a dynamic adaptive filter to calculate and subtract a synthetic echo from the return signal. It allows double talking.

with their many adjustments, switching delays, and other problems are far from ideal. The problems of echo suppressors are avoided with *echo cancellers*.

Echo cancellers create a synthetic echo that is subtracted from the return signal, thereby cancelling the echo, as shown in Figure 3.38. Because different local loops have different electrical characteristics, the precise echo cannot be known in advance. The cancellation is therefore performed with electronic filters that can change their properties dynamically to minimze any portion of the returned signal that resembles the incoming speech signal. Echo cancellers have become affordable through the use of integrated circuits and digital processing. A short amount of time is needed by them to adapt to the properties of the connection, and a careful listener can hear the quick disappearance of the echo at the very beginning of the call. Echo cancellers allow two-way talking and full-duplex operation on long-distance calls.

TRANSMISSION IMPAIRMENTS

The quality of transmission can be impaired by a variety of sources. These sources all produce unwanted signals, which can be grouped under the general category of *noise.*

Amplifiers generate noise in the form of *hiss.* Because the frequency spectrum of this noise is uniform with all frequencies equally present, like the spectrum of white light, such noise is called *white noise* by analogy. Motors turning on and off, electromagnetic disturbances produced by lightning, and other such sources can generate *clicks* and *pops.* These clicks and pops are called *impulse noise* because the noise looks like sharp spikes or impulses. Loose connections can produce a *sputtering noise.* The carbon microphone is a source of *granular noise.* The quantization used with digital conversion of the speech signal results in a small error upon reconstruction of the analog signal, and this error appears in the form of *quantization noise.*

Crosstalk is the leakage of one signal into another circuit. It can occur in the local loop where many lines all run parallel to each other. Crosstalk can also occur on long-distance circuits using frequency-division multiplexing when the signal from one channel leaks into another because the frequency shifting drifts a little or because of other causes.

Transmission systems cannot be perfect in the sense of eliminating all forms of impairments—that would be far too costly. Instead, the design goal is to reduce their subjective effect on users to an acceptable level. Hence, the human factors of transmission impairments are an important consideration in the performance specifications for transmission systems.

TRANSMISSION—AN ASSESSMENT

Transmission has clearly become very inexpensive, thanks to the vast capacity of optical fiber along with the ease of time-division multiplexing. AT&T no longer has a monopoly on long-distance service, and many carriers have entered the long-distance business with their own networks. With all these entrants and all this capacity, a glut in circuits has occurred, and long-distance rates will continue to decrease. Competition between the various carriers will become more intense because in this world of digital, one carrier's bit is the same as any other carrier's bit. Differences in rates, ancillary services such as operator service, and maintenance quality will be stressed to distinguish one carrier from another.

As rates continue to decrease, one wonders where it will all lead. Will residential traffic increase? Will people talk longer during social calls? Will facsimile calls substitute for mailed letters? How much time can people spend on the telephone?

Although technology has had great impact on long-distance transmission, the local loop remains old-fashioned, simple, twisted pairs of copper wire. The local loop with its restricted bandwidth and capacity is viewed as a communication bottleneck—but is it?

The key to communication capacity is understanding communication needs. Most communication services can be transmitted quite easily over the existing local loop. Voice communication is already adequately served, and, in fact, the residential local loop is idle most of the day. Most data services rarely require more than 4800 bps, and this rate can be easily carried over the existing local loop. Only one service could tax the local loop, and that service is two-way video communication—the *picturephone* or *videophone*. However, the consumer need for the picturephone is highly suspect, and many people simply do not want to be seen while talking over the telephone. Some persons are so negative about being seen while speaking over the telephone that they would probably pay extra *not* to have a picturephone! More is said about these and other services in a later chapter.

REFERENCES

"AR6A Single-Sideband Microwave Radio System," *Bell Syst. Tech. J.,* Vol. 62, No. 10, Part 3, December 1983.

Elbert, Bruce R., *Introduction to Satellite Communication,* Norwood, MA: Artech House, Inc., 1987.

Hudson, Heather, *Communication Satellites,* New York: The Free Press, 1990.

Keiser, Bernhard E., and Eugene Strange, *Digital Telephony and Network Integration,* New York: Van Nostrand Reinhold Company, 1985.

Nagel, Suzanne R., "Optical Fiber—The Expanding Medium," *IEEE Comm. Mag.,* Vol. 25, No. 4, April 1987, pp. 33–43.

O'Neill, E.F., ed., *A History of Engineering and Science in the Bell System: Transmission Technology (1925–1975),* Indianapolis, IN: AT&T Bell Laboratories, 1985.

Talley, David, *Basic Carrier Telephony,* Rochelle Park, NY: Hayden, 1977.

Chapter 4
SWITCHING

EVOLUTION

In Bell's invention and first demonstration, telephone communication went from one instrument to another a few rooms away. This type of *private-line service* characterized initial telephone installations: one telephone could communicate with only one other telephone—not a particularly exciting form of communication. Early on, a number of telephones were connected to one line, a form of *party-line,* so that a number of people could participate in the conversation, as shown in Figure 4.1. But there was no way to reach telephones connected to any other line. The solution was to find a means by which any telephone could be connected, or *switched,* to any other telephone.

One way, though impractical, of accomplishing such switching would be to bring pairs of wires, or lines, from all telephones to all other telephones. In this way, each telephone would simply connect by a simple switch to the line to the desired party's telephone, a form of *station switching.* This would be fine if only a small number of lines were involved, but if hundreds of telephones needed to be reached, then hundreds of lines would need to go to each telephone, clearly an impractical arrangement.

The practical solution was discovered quite early and was the implementation of a *centralized switching* arrangement. All the lines from all the telephone stations were brought to a common place, a *central office,* where the electrical cross connections could be made to connect one station to another. The central office was also called the telephone *exchange.* The actual connections were made by human operators who operated the exchange.

As exchanges grew to cover larger geographical areas, it became uneconomical to bring all the lines from outlying areas to a single central office. More central offices were created, each serving a nearby surrounding area. A call from a party served by one office to a party served by another office was completed over dedicated lines, called *trunks,* that interconnected the offices. The operator in one office reached an operator in the other office over a trunk circuit and relayed the identity

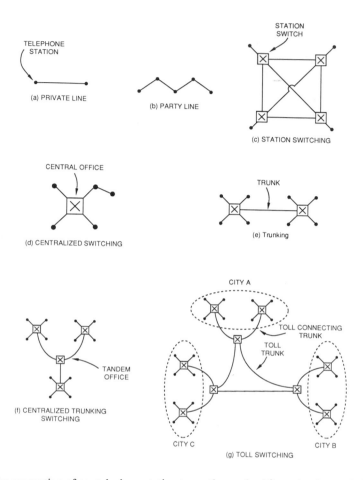

Figure 4.1 The connection of one telephone station to another evolved from simple unswitched private-line service to complex intercity networks involving many stages of switching. Station switching (c) is impractical since lines from all other stations must be brought to each and every station. Instead, lines are brought to a central office (d), and calls involving a second office are completed over trunks (e). Tandem offices (f) serve a number of central offices, and toll offices (g) switch calls from city to city.

of the called party so that the distant operator could complete the call. As growth continued, special switching offices, called *tandem offices,* were developed to handle the switching of interoffice calls carried over trunks between local offices.

Extended switching systems were developed to serve long-distance and toll circuits between distant cities. Switching offices, called *toll offices,* were developed that performed switching solely for toll circuits.

For many businesses, most telephone traffic was between telephones that were all located on the customer's premises. This traffic could be served most efficiently by a private switching system located on the premises. This switch was called a *private branch exchange,* or a PBX. Today's PBXs are automatic (PABX), mostly using electronic technology (EPABX).

TELEPHONE NETWORK SWITCHING

Switching Offices

The telephone network is composed of a number of centralized locations, called *switching offices,* where the telephone conversation carried on one circuit is switched, or connected, to another circuit. The three major types of switching offices are *local, tandem,* and *toll.* The local office, sometimes called the *end office* or *serving office,* is the one closest to the telephone station and connects directly to the local loop. A tandem office serves a cluster of local offices. Toll offices, or toll centers as they are sometimes called, are involved in the switching of traffic over long-distance, or toll, circuits.

Switching Hierarchy

In the past, switching offices were organized into a hierarchy with the end offices where the call originates and terminates being at the lowest level. The end office is termed a class 5 office in the hierarchy. The end office connects to a toll center (a class 4 office); the toll center connects to a primary center (a class 3 office); a primary center connects to a sectional center (a class 2 office); and finally at the very top of the hierarchical pyramid, a sectional center connects to a regional center (class 1 office). This hierarchical structure is depicted in Figure 4.2. While there are well over 20,000 end offices, there are only about 20 regional centers serving the United States and Canada.

Normally a call is completed using as few switching offices as possible. An end office will attempt to make the connection directly to the terminating end office over a direct interoffice trunk. If necessary, the switching capabilities of a tandem office will be used. If the call is a toll call, then the toll network and its toll switching offices become involved.

With the hierarchical system, the preferred route was to connect across from an originating office to a lower terminating office using routes called *high-usage routes.* If this route were busy, then the next preferred route was to connect across to a center at the same level, or if this route were also busy, then across to the next higher center. As a last resort if all the routes across were busy, then the call would ascend to the next higher originating office and attempt to cross over again.

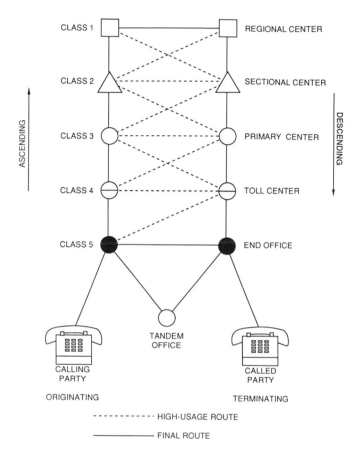

Figure 4.2 This switching hierarchy was used in the past. Present telephone network switching is a dynamic nonhierarchical system in which all the toll switching centers are equal, and the route for each call is chosen according to available capacity.

The routing used in today's long-distance networks is a dynamic nonhierarchical system in which all toll switching centers are equal, and the route for each call is chosen according to the capacity that is available. Thus, with this newer system, the final routes are not fixed but are chosen dynamically at the moment of the call. Increased flexibility and efficiency are possible with this dynamic nonhierarchical system.

Local Office

The local central office is where the initial and terminating switching occur. As shown in Figure 4.3, all the local loop wire pairs enter the central office at the *cable*

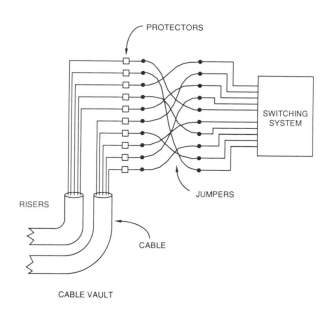

Figure 4.3 Jumpers connect from the protectors to the actual input to the switching equipment in a central office.

vault, usually located in the basement of the building. The cables run in all four directions of the compass and usually enter the building from conduits. The cables then turn upward through *risers* to upper floors of the building. A photograph of a cable vault is shown in Figure 4.4. Each individual wire pair has its own *protector,* which is a small circuit breaker-fuse designed to protect the sensitive and costly switching equipment from excessive electric voltages and currents. All the protectors are situated in a *protector frame,* shown in Figure 4.5. Wires called *jumpers* interconnect from the protector frame to the actual inputs to the switching system. All these jumper wires are situated at the *main distribution frame* (MDF), shown in Figure 4.6. The jumble of long wires in older MDFs is minimized in the newer AT&T COSMIC® frame (for Common Systems Main Interconnect), shown in Figure 4.7.

When I visit a central office, I am amazed, in this day of modern electronics, to see such large amounts of wire. When I descend into the cable vault, I see that much of the cable is thick and black. However, I also see much thinner orange cable. This represents optical fiber that is being used for trunks to other offices and to interexchange carriers. Fiber is also being used in feeder systems in the local loop. When I visit the upper floors of the office, I am surprised at all the empty space. That space was formally occupied by old, electromechanical switching equipment, which was probably removed during the past 10 years or so. If I am very lucky, I might still see an old crossbar switching system and obtain an intuitive feel for the

Figure 4.4 Photograph of a cable vault. Cables enter the vault horizontally from underground conduits and then turn upward toward the floors containing the switching equipment.

mechanics of switching that is impossible to obtain with today's electronic systems. Switching technology has evolved greatly over the years.

TECHNOLOGY EVOLUTION

During the first few decades of telephone service, switching was a manual operation performed by human beings, called *operators,* who made the actual connection of circuits. (Refer to Figure 4.8.) The connections were made at a switchboard utilizing cords with plugs at the ends. The plug had a tip and a ring, which completed the actual electric circuit over which the telephone signals traveled. The sleeve was used for signaling and supervisory purposes. The terms "tip" and "ring" continue to this day for the two wires between the central office and the telephone instrument.

Figure 4.5 Photograph of a protector frame. Each line has its own protector to protect the switching equipment from excessive voltages and currents.

Each operator had about 18 cords that could be used to make connections. About 120 incoming lines terminated at the operator's switchboard at answering jacks. A lamp at each jack would light indicating that the line desired service. The operator would plug in and answer the call. The calling party would give the name, and later the telephone number, of the called party. The operator would then complete the call by connecting the cord to one of perhaps as many as 10,000 subscriber

Figure 4.6 Photograph of a main distribution frame (MDF). Here jumper wires connect the protected lines to the switching machine.

multiple jacks within reach. Lamps associated with each connecting cord indicated when parties had completed a call.

Manual switching was slow and very costly in terms of its extensive use of human labor, as shown by the photograph of Figure 4.9. However, the use of humans gave it considerable intelligence and a friendly (we hope) interface with users.

The first major innovation in switching came in 1892 with the first installation of an automatic switch controlled by the subscriber's telephone instrument. The switch was conceived by Almon B. Strowger. A later modification of his invention included the invention of the rotary dial and the use of dial pulses to control the operation of the switch. The Strowger switch was an electromechanical device that was later adapted for use in the Bell System starting in 1919. Bell System engineers were able to develop improved electromechanical switching systems beyond Strowger switching. However, *electromechanical switching systems* were slow and noisy and were not very flexible in terms of offering new services. But they most

Figure 4.7 Photograph of an AT&T COSMIC® mainframe. The jumble of wires of a conventional frame is avoided through careful design.

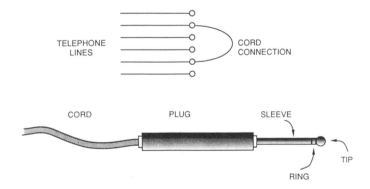

Figure 4.8 With manual switching, a three-conductor connecting cord makes the electric connection from one telephone line to another. The cord terminates at each end in a plug in which the metallic transmission circuit is formed by the tip and ring of the plug. The sleeve is used to complete signaling circuits used within the central office for such purposes as indicating that a circuit is in use.

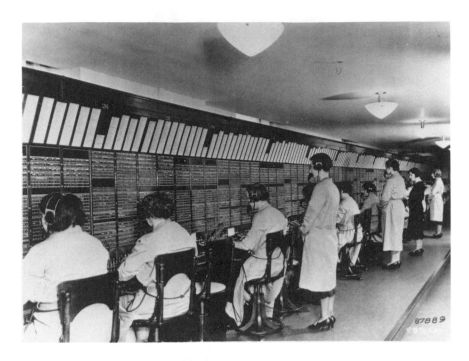

Figure 4.9 Photograph of a manual switchboard capable of serving 5000 lines. (Courtesy of AT&T Bell Labs.)

certainly were less costly than human operators and were directly repsonsible for making telephone service affordable and universal.

Today's telephone switching systems are *electronic switching systems* using either analog space-division switching or digital time-division switching. The electronic technology is fast, and the programmable computer that controls the switches offers considerable flexibility.

SWITCHING SYSTEM

There are two functional parts to any telecommunication switching system, as shown in Figure 4.10. The first part performs the actual switching and connection of one circuit to another. This part consists of the *switches,* which are interconnected to create a network of switches, commonly called the *switching network.* The second part of a telecommunication switching system is responsible for issuing the commands necessary to operate the switches at the proper time to make specific connections. This part is called the control function, or simply *control.*

Various switching systems have been designed and developed over the years. The switching network and the control in the earliest systems, such as the step-by-

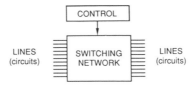

Figure 4.10 A telephone switching system consists of two major parts: the switching network itself, which is the actual means for connecting one circuit to another, and a means of control for instructing the switching network to make specific connections.

step system and the crossbar system, were completely electromechanical. Newer systems utilized electromechanical switching networks but electronic control in the form of a programmable digital computer; these systems were called *electronic switching systems*. The newest switching systems are totally electronic having both an electronic switching network and electronic control.

We will examine the progression of switching technology through study of the workings of these various systems. But first we need to examine, from a theoretical perspective, the various ways in which switching is accomplished. Then we will describe specific switching systems that have been developed and used over the years, starting with the earliest electromechanical systems and ending with today's digital systems.

APPROACHES TO SWITCHING

There are two basic ways in which the switching performed by the switching network can be accomplished: *space-division switching* and *time-division switching.*

In space-division switching, as Figure 4.11 shows, each telephone conversation has its own dedicated physical path through the switching network, and the

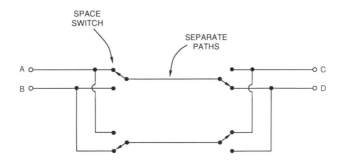

Figure 4.11 A simple space-division switching system utilizes two space switches to enable two circuits to be connected to two other circuits. Each electrical connection is established over separate physical paths through the switching network. In this example, party *A* is connected to party *D*, and *B* is connected to *C*.

physical connection is maintained for the duration of the call. Switches that switch signals in actual space over separate electrical paths in the switching network are used in space-division switching systems. There are many physically separate electrical paths in a space-division switching system.

In time-division switching, as shown in Figure 4.12, separate conversations share the same electrical path through the switching system but are separated in time. In essence, the path is time shared. Each conversation is sampled, and the sample values are packed together over the shared path until they are separated and routed to the appropriate destination. Clearly, time-division switching is most appropriate for signals that have been sampled and digitized. For this reason, time-division switching systems are commonly called *digital switches.*

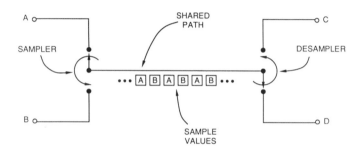

Figure 4.12 In space-division switching, a path is shared by many circuits for very short time intervals. In this example, the signal on circuit *A* is sampled by a sampler switch, and these sample values are interspersed between sample values from circuit *B* for transmission over a common, shared path. At the other end, the sample values are momentarily connected to circuits *C* and *D* in an order such that samples from circuit *A* go to circuit *D* and samples from circuit *B* go to circuit *C*.

With space-division switching, signal paths are switched in physical space. With time-division switching, sample values of a number of signals sharing a common path are reorganized, or switched, in their time sequence.

Space-division switching is usually associated with the world of analog and time-division switching with the world of digital. However, the switching universe is not quite that simple, and modern digital time-division switching systems usually utilize switches that switch signals both in space and in time. All this will become clearer later in this chapter when time-division switching is explained in detail.

Quite clearly, space-division switching is somewhat akin to space-division multiplexing, and likewise time-division switching is akin to time-division multiplexing. Hence, one might wonder whether there is some form of switching that is akin to frequency-division multiplexing, namely, *frequency-division switching.* With frequency-division switching, different frequency channels are assigned to different connections. This is the way cellular telephony works, as is explained in a later chapter. Each telephone conversation uses a specific frequency channel in the radio spectrum. This type of switching is also particularly applicable to a single broadband medium shared by a number of users, such as the coaxial cable of CATV.

SPACE-DIVISION SWITCHING

Basic Switches

The electromechanical switch invented by Strowger was a stepped rotary switch that moved in two dimensions. It will be described in more detail in a following section on the step-by-step switching system. In its most general form, a rotary switch connects one line to any one of many lines, as shown in Figure 4.13. The switch rotates through its positions to make the final desired electrical connection. Though simple in concept, its fault is that electrical contacts must slide across each other, thereby subjecting them to considerable wear, and wear creates noise. Furthermore, the relatively large amount of mechanical motion means that a fair amount of time is needed to make an actual connection.

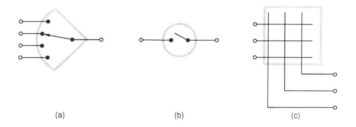

(a)   (b)   (c)

Figure 4.13 The basic types of switches used in space-division switching are (a) the rotary switch, (b) the on-off switch, and (c) the matrix switch.

We are all familiar with the on-off switch that turns lamps on or off. In its electromechanical form, it consists only of two contacts and a means to complete or to open the circuit between them. Though simple, the electromechanical on-off switch is still subject to some wear, and though faster than a rotary switch, it is still somewhat slow in today's world of electronics. A transistor can be used as an on-off switch and offers the advantages of speed, low noise, and no wear.

On-off switches are used to create a basic *coordinate,* or *matrix* switch. A matrix switch can connect many lines with many other lines. Connections are made with on-off switches at the crosspoints in a matrix of all input lines and all output lines. The electrical connection at each crosspoint can be made using a variety of electromechanical and electronic technologies, such as conventional contacts, small reed contacts sealed in glass, and diodes and transistors.

Switching Stages

Normally, not everyone will want to converse with everyone else at the same time. Hence, a switching network does not need to be designed so that all incoming lines can be connected to all outgoing lines. The incoming lines can therefore be concentrated and distributed over a smaller number of switching paths before being expanded at the last stages of switching, as shown in Figure 4.14. In this fashion, the switching is accomplished in stages consisting of *concentration, distribution,* and *expansion.*

Figure 4.14 There are three stages of switching in a space-division switching network. Many input lines are concentrated together into a smaller number of serving lines. The ratio of the number of input lines to serving lines is greater than one. These serving lines are then distributed to output switches, which expand service to a large number of output lines. The ratio of the number of output lines to input serving lines is greater than one.

A basic three-stage switching network using rotary switches is shown in Figure 4.15. The electromechanical Strowger rotary switch is able to connect one line to any one of 100 other lines and is the basic technology used in the switching network in the step-by-step switching system. The step-by-step system is explained in a following section.

A basic three-stage switching network using matrix switches is shown in Figure 4.16. The electromechanical *crossbar switch* is a matrix switch that can connect

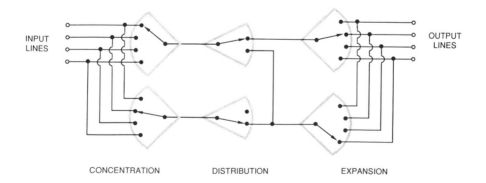

Figure 4.15 A basic rotary three-stage switching network. If more than two input lines demand service, they will be denied service because there are only two input switches available to serve them.

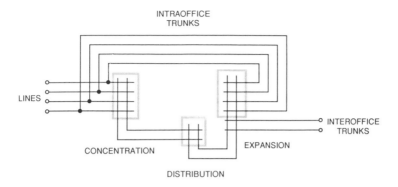

Figure 4.16 A basic matrix, or coordinate, three-stage switching network. Electric on-off switches make the required connections at the crosspoints. Calls within the office are made over intraoffice trunks that connect the output from the switching network back to the input lines.

10 inputs to 20 outputs. The crossbar switch, invented at Bell Labs, is used in the switching network in crossbar switching systems. The crossbar system is explained in a later section.

Many telephone calls are between lines served by the same switching office. To handle these calls, the output from the switching network is connected back on itself to the incoming lines to make these intraoffice connections. The lines used for such intraoffice calls are called *intraoffice trunks*.

BLOCKING

Clearly, it is possible for some paths in the switching network to become completely congested so that no more calls can be carried. In this case, any additional calls desiring service will be *blocked*. The blocking can occur at any stage, and the switching network must be designed to minimize blocking for the maximum traffic that it handles. The acceptable amount of blocking depends on the type of customer served by the switching office, insofar as different types of customers have different traffic patterns. In particular, business customers usually make many more calls during the day than residential customers.

It is possible to design a switching network so that it is nonblocking. However, such a design usually takes a large number of switches, and switches are costly. Through appropriate design, it is usually possible to have a switching network with an acceptable amount of blocking and a fewer number of switches than for a nonblocking network. The example of Figure 4.17 shows a 12-by-12 matrix switch serving 12 input lines and 12 output lines. This matrix switch, requiring 144 on-off switches, is strictly nonblocking, and any input can be connected to any output. A three-stage switching network can be more efficient, particularly if some amount of blocking is acceptable. Only 84 on-off switches are needed in the three-stage block-

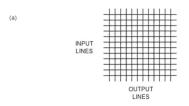

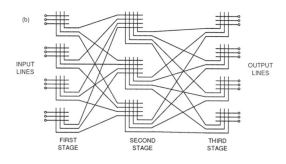

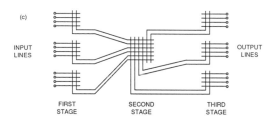

Figure 4.17 (a) A 12-by-12 matrix serves 12 input lines and 12 output lines requiring 144 on-off switches at the crosspoints. This matrix switch is strictly nonblocking. (b) Through efficient design, the number of switches can be reduced with a three-stage switching network to 120 on-off switches. However, although the switching network is nonblocking, the switch contacts might need to be rearranged to prevent blocking in some cases. (c) If some blocking is acceptable, the switching network can be even more efficiently designed, this example requiring only 84 on-off switches.

ing design. If rearranging the switch contacts during calls is acceptable, a nonblocking design requiring 120 on-off switches is possible.

Central office switching systems typically serve 10,000 lines. A strictly nonblocking matrix switch would require 100 million on-off switches and would be prohibitively costly. Because not all lines normally require service at the same time, more efficient multistage designs with acceptable amounts of blocking are used. Of course, there are those rare times when everyone does demand service, such as in a disaster, and many customers will have to be turned away.

TIME-DIVISION SWITCHING

Time-Slot Interchange

In the chapter on transmission, we studied how a number of voice signals can be multiplexed onto a single transmission path using time-division multiplexing. Each voice signal is converted to a digital format, and each digital circuit is then assigned its own unique time slot in the stream of digital information. To switch these circuits, it is only necessary to interchange the digital signals in the time slots.

Consider a transmission path carrying the signals from four digital voice circuits. The digitized sample values, A1, A2, A3, and A4, corresponding to the four digital circuits flow sequentially along the transmission path in the order A1-A2-A3-A4, and so forth, as shown in Figure 4.18. The 8 bits of digital information, A1, in the first time slot correspond to the first circuit, the second time slot corresponds to the second circuit, and so on. This stream of digital information forms the input to a time-division switch. The output from the switch likewise is a stream of digital information. However, the sequence has been reordered by the switch. The first time slot now contains A2, the second time slot contains A4, the third time slot contains A3, and the fourth time slot contains A1.

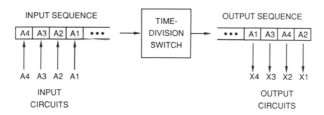

Figure 4.18 Time-division switching is accomplished simply by reordering the sequence of information contained in the time slots in a digital transmission path. In this example, input circuit A1 is connected to output circuit X4; A2 to X1; A3 to X3; and A4 to X2. This technique is called time-slot interchange.

The information in the first time slot in the output sequence is connected to the first output circuit, X1; the second time slot is connected to the second output circuit, X2; and so forth. Thus, in this example, since the first time slot in the output sequence contains A2, the second input circuit has in effect been connected to the first output circuit, X1. Similarly, A4 has been connected to X2; A3 to X3; and A1 to X4.

In this example, switching has been accomplished through the simple interchange of information contained in time slots. This is called time-division switching, and it is performed by a technique called *time-slot interchange*. Since the signals switched in a time-division switching system are digital, time-division switching systems are sometimes called *digital switches.*

Time-slot interchange is performed through the use of a temporary storage called a *buffer memory*. The information contained in the input time slots is entered in sequence into the buffer memory, as shown in Figure 4.19. The information is then read from the buffer memory in a different sequence corresponding to the desired order of switching.

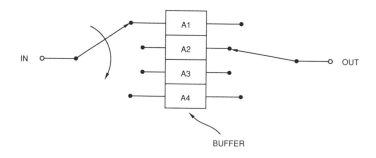

Figure 4.19 Time-slot interchange is performed through the use of a buffer memory into which the input samples are entered in the sequence received. The samples are then read from memory in a different sequence by a programmable switch. The order of the output switch in this example is 2-4-3-1, thereby reordering the output samples to correspond to the switching of Figure 4.18.

It takes time to write information into memory and then read the information from memory. This amount of time is called the *write-read memory cycle time,* and it determines the maximum number of time slots that can be switched in the buffer. The information in a particular location in the buffer usually waits awhile before being read out. This waiting time contributes a delay to the signal. The maximum delay is the total length in time of all the time slots stored in the buffer. Because the time slots for a full frame are usually stored in the buffer, the maximum delay is the length of a frame, or 125 μs.

Normally, a large number of digital circuits are involved in a telecommunication system. These digital circuits come from digital transmission trunks onto which a number of digital circuits have been multiplexed through time-division multiplexing. Or the digital circuits can come from a number of analog voice circuits that have been digitized and multiplexed onto a digital line solely for the purpose of being switched in a time-division switching system. Either way, usually a number of physically separate multiplex lines are involved. Because a circuit on one multiplex line might need to be switched to a circuit on another multiplex line, some form of physical switching between different digital multiplex lines is needed. Such physical switching is accomplished using a space switch, and hence time-division switching systems usually involve a combination of both space switches and time switches.

The space switch used in a time-division switching system is usually a matrix switch that must continuously alter its specific connections according to which time slots from different physical lines must be switched to other lines. Because this space switch is being continuously switched in time and shared by many lines, it is truly a time-multiplexed space switch and is called a *multiplexed space switch*. The following example explains the operation of a digital switching system operating in three stages: time, space, and time.

Example of a Digital TST Switching System

The switching networks used in most digital switching systems use both time-division switching and space-division switching. The sequence of time switches (T) and space switches (S) categorizes the switching system. The following example has three stages of switching in the following order: time-division switching (T), space-division switching (S), and time-division switching (T). This type of switching system is called a *TST switching network.*

An example, adapted from Hills (1978), should clarify these concepts. Refer to Figure 4.20. Two digital trunks, A and B, each carrying four digital circuits are to be connected to two other digital trunks, X and Y, likewise each carrying four digital circuits. The connections are to be made as follows:

A1-X4   B1-X1
A2-Y3   B2-Y1
A3-Y4   B3-Y2
A4-X2   B4-X3

The input digital trunk lines A and B are each connected to time-slot interchange units. The digital samples from each line are entered into the two buffers in the sequence received. The output digital trunk lines X and Y are likewise connected to two time-slot interchange units, and the samples for each output line are read from the buffers in sequence from the first location in memory to the last.

A multiplexed space switch with two inputs and two outputs connects the four time-slot interchange units. The space switch continuously reconfigures itself to send the appropriate sample values in the input buffers to the appropriate locations in the output buffers. The input data are arriving at a rate of 4 time slots every 1/8000 second, or 32,000 time slots per second. Thus, the multiplexed space switch needs to reconfigure itself 32,000 times per second. The *switch time* is the time between each reconfiguring, and in this example the switch time is 1/32,000 second, or 31.25 μs.

The information stored in the input time-slot interchange units (TSIUs) is read from the buffers in an order keyed to the switch order of the space switch.

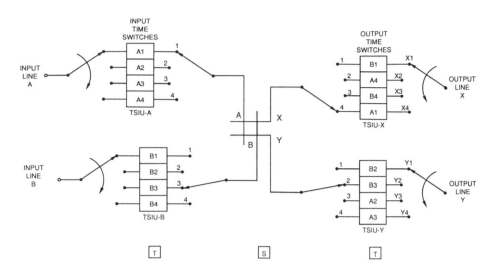

Figure 4.20 This digital switching network connects the four circuits in each of two input lines to four circuits in each of two output lines. The switching is accomplished in three stages: time-division switching, space-division switching, and time-division switching. [This example is from M.T. Hills, *Telecommunications Switching Systems,* Cambridge, MA: MIT Press, 1978, p. 289 (Figure 12.1).]

Similarly, the information entered into the output TSIUs is keyed to the operation of the space switch. The whole switching network must perform in perfect synchrony according to a master switching plan. The switching plan for the simple TST system is shown in Figure 4.21. For example, at switch time 2, the information contained in time slot A4 of the TSIU serving line A is switched by the space switch to the TSIU serving output line X. That TSIU stores the information in the second slot for later connection to X2. Thus, A4 has been switched to X2.

Clearly, even for such a simple example, this is a complicated process, requiring a precise switching order for all the switches in the network. This switching order is determined by the computer controlling the switching system in response

SWITCH TIME	SPACE SWITCH		TIME SWITCH			
	A	B	A	B	X	Y
1	X	Y	1	3	4	2
2	X	Y	4	2	2	1
3	Y	X	3	4	3	4
4	Y	X	2	1	1	3

Figure 4.21 The various time and space switches in the switching network of Figure 4.20 operate according to this switching order.

to the switching needs of the various circuits. Because all input circuits can be connected simultaneously to all output circuits, the TST switching network in this example is a nonblocking switch.

Other Approaches

Large buffers were costly in the past, and also the write-read cycle time was somewhat long. Thus, large buffers were avoided through the use of multiplexed space switches to connect digital circuits on different input and output lines. However, storage continues to decline in cost, and write-read cycle times likewise continue to decline. Therefore, large buffers and fewer time-multiplexed space switches will be used in future digital switching systems.

The preceding example used a combination of time and space switching to accomplish the switching. The same switching could be accomplished through the use of two large buffers, as shown in Figure 4.22. The input samples from the two input trunks are entered in parallel into both buffers. Each buffer then serves one output trunk. An alternative approach, assuming a buffer memory that could have multiple, simultaneous inputs and outputs, would be the use of a single buffer memory, as shown in Figure 4.22(b).

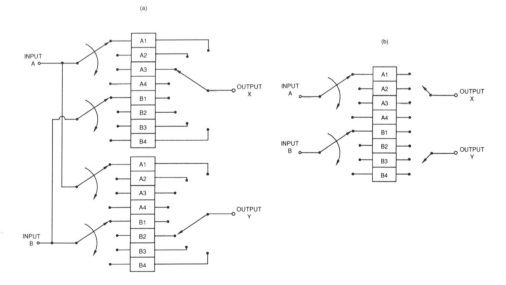

Figure 4.22 The time-multiplexed space switch can be avoided through the use of larger buffer memories. (a) Two buffers are used for the two output lines, and all the input time slots are entered in parallel into both. (b) A single buffer with dual-access inputs and outputs accomplishes the same switching.

Specific Switching Systems

The two major technologies used in switching systems are electromechanical and electronic. Electromechanical is the way of the past; nearly all modern switching systems today are electronic.

Electromechanical switches are the foundation of space-division switching systems. The earliest type of electromechanical switching system was the *step-by-step* system using the Strowger rotary switch. Another type was the *panel system,* although it was relatively short-lived. The next type was the *crossbar system,* which utilized a coordinate or matrix switch. Electromechanical control technology was used in the step-by-step, panel, and crossbar switching systems. This technology was permamently hard-wired, and features and functionality could not be changed easily. Electromechanical control was nonprogrammable and could not be reprogrammed.

The solution to the inflexibility of electromechanical control was the use of programmable, electronic, digital computer technology for control of the switching system. Switching systems using such control were called *electronic switching systems* (ESS) in the Bell System. The Bell System's No. 1 ESS® system used electronic control, although the switching network was a space-division system, initially using electromechanical switch contacts. Later, electronic diodes were used in the switch matrices.

Modern switching systems use time-division switching for the switching network along with computer control. The control can be located centrally, as in the AT&T No. 4 ESS® system, or distributed throughout the switching system, as in the AT&T 5ESS® switching system. The DMS® family of digital switching systems manufactured by Northern Telecom uses a combination of distributed and central control.

These specific systems are explained in the following sections. Other vendors manufacture digital switching systems, but the AT&T and Northern Telecom machines are most prevalent in North America, and hence these machines are featured.

THE STEP-BY-STEP SWITCHING SYSTEM

History and Invention

The basic switch used in the step-by-step switching system was invented by Almon B. Strowger, an undertaker. The anecdotal story is that Strowger invented the switch because he was upset that all the business in his town went to his competitor, whose wife was the operator of the local exchange. His invention was an automatic switch that stepped vertically and horizontally under the control of digits entered at the telephone instrument. His switching system was first installed in La Porte, Indiana, in 1892. A few years later, in 1896, the telephone dial was invented by some of Strowger's associates. The Strowger switching system was manufactured by the Automatic Electric Company, formed in 1901, and was sold primarily to the non-Bell, independent telephone companies. The first installation of a Strowger switching system in the Bell system did not occur until 1919. Until then, the Bell companies had continued to use human operators and had ignored the new technology for nearly 25 years. The Strowger system was called step-by-step in the Bell System.

As late as the late 1980s, it was still possible to find step-by-step switching systems in use in the Bell System and in independent companies. However, step-by-step switching is virtually extinct today in the United States. Step-by-step switching had a lifetime of nearly 100 years, which clearly indicates the reliability and utility of Strowger's invention. It is very rare indeed for an electromechanical technology to have such a long lifetime.

The Strowger Switch

The key ingredient in the step-by-step switching system is the basic Strowger switch, shown in Figure 4.23. The Strowger switch is an ingenious, electromechanical device formed from electromagnets and ratchets. The electromagnets are activated by the dial pulses and ratchets cause a wiper assembly to step vertically and then to rotate horizontally in steps until the desired contact is reached (see Figure 4.24). There are 100 contacts arranged in a bank of 10 rows by 10 columns.

There are two banks of contacts in a Strowger switch. The bottom bank is called the *line bank,* and it is where the actual tip and ring connection is made for the telephone conversation. The top bank is called the *sleeve bank,* and it is where the control connection is made. A direct current flows through the sleeve contacts to maintain an actual connection. When the calling party hangs up, the flow of current through the sleeve circuit ceases, and a spring causes the switch to return to its resting position.

There are a number of relays at the very top of a Strowger switch. These relays count the dial pulses and control the electromagnets that cause the wiper to move.

The action of the wiper as it slides across the contacts results in considerable wear of the contacts, which introduces a fair amount of electrical noise into the connection. There are many mechanical adjustments that need to be made to keep the switch working properly. Maintenance is a nearly continual process. Nevertheless, the basic simplicity and high reliability of the Strowger switch led to nearly a century of use.

Figure 4.23 Photograph of a basic Strowger switch. The contact banks and wiper arm are shown along with the relays at the top that control the movement of the wiper arm.

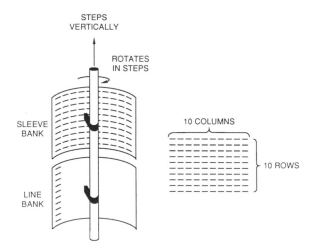

Figure 4.24 A wiper arm moves vertically in steps to reach the desired row of contacts. It then rotates horizontally in steps until the exact column is reached, thereby making the desired connection.

Switching System

A step-by-step switching system usually serves 10,000 telephone lines. A typical step-by-step switching system is a very large machine requiring many bays of equipment, as shown in Figures 4.25 and 4.26.

We next describe the operation of a step-by-step switching system that can reach any one of 10,000 lines, numbered 0000 through 9999, by the dialing of a four-digit telephone number. The dialed number corresponds exactly to the location of the contacts of the called party's line in the switching system.

Figure 4.25 A step-by-step switching system is composed of many frames of Strowger switches. The clicking noises of these switches can reach a crescendo during busy periods.

Figure 4.26 A considerable amount of wire interconnects the rear of the Strowger switches in a step-by-step switching system.

The switching network is divided into stages, illustrated in Figure 4.27. The first stage is the *line finder*. When the calling party lifts the telephone handset from its cradle, the telephone instrument starts to draw direct current from the telephone line. The initiation of this flow of dc causes relays in the switching system to initiate a hunt by the line finder to locate the particular line that is drawing current and thus wishes service. Each line-finder switch serves 100 lines. Once a particular line-finder switch finds a line and connects to it, that line-finder switch is engaged and

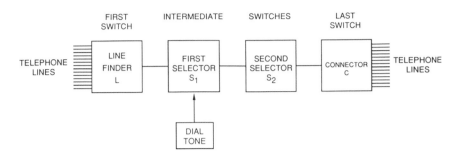

Figure 4.27 The switching network in a step-by-step system designed to reach any one of 10,000 telephone lines is accomplished in stages. The line finder searches for the direct current flowing in the line desiring service. The first selector returns dial tone and reponds to the first digit. The second selector responds to the second digit. The last two digits operate the connector, which makes the final connection to the dialed party's line.

can no longer serve any other lines. Therefore, a number of line-finder switches are connected in parallel to serve the same 100 lines. The specific number of line-finder switches depends on the anticipated traffic and usually is 10 or less.

Each line-finder switch is hard wired to the next switch in the sequence of switching stages. This next stage is called the *first selector*. The first selector connects the calling party to dial tone and then waits for dialing to commence. The first dialed digit causes the first-selector switch to move vertically a number of steps corresponding to the dialed digit. The switch senses the pause between dialed digits and then searches, or hunts, horizontally for an idle line to a second selector switch.

The second-selector switch moves vertically with the second dialed digit and then hunts horizontally for an idle line to the last switch in the process. If either selector switch cannot find an idle line, the wiper passes by the tenth column of contacts and reaches an eleventh contact, which returns a fast-busy tone to the calling party.

The last switch is called a *connector*. It first moves vertically with the third dialed digit and then rotates horizontally with the fourth dialed digit. The connector switch makes the final connection to the number dialed.

The above-described switching system can connect to any one of 10,000 lines in a single switching office. Other offices in a local exchange area are assigned three digits to indicate the specific office and switching system. These three office digits are then followed by the four-digit number of the line within the office. The calling party thus dials seven digits. Three additional selector switches are added to the step-by-step system to accommodate the office digits. If the call were to another office, the third selector switch would connect to an interoffice trunk to that office, and the last four dialed digits would make the final connection at the distant office.

Control

The control function in a step-by-step switching system is distributed throughout the system in the form of the relays that control each individual Strowger switch, as shown in Figure 4.28. There is no central control in a step-by-step system; each switch has its own control associated with and incorporated into it. The dial-pulse digits contain the information necessary to operate each switch. These dial pulses are carried over the same lines that will carry the speech signal. The control path thus is the same as the speech path.

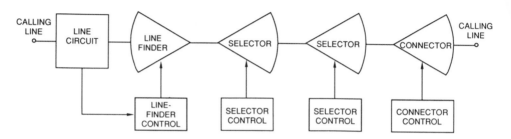

Figure 4.28 The control in a step-by-step switching system is distributed throughout the system in the form of the relays that control each individual Strowger switch. Each dialed digit directly controls the movement of each Strowger switch as the path through the switching network progresses. This type of control is called *direct progressive control.*

The earliest step-by-step systems used a method of control called *direct progressive control.* With this method of control, each stage of switching responds directly to the digits dialed by the calling party. As one stage completes its connection, the next stage responds progressively to the next dialed digit. The connection is completed progressively through the switching network.

Direct progressive control is simple, but it has two problems. Because the paths through the switching network are created progressively, there is no way to look ahead to determine whether future paths are already in use. Thus, a call might be blocked because no idle paths are available. This means that additional spare capacity is needed to minimize the effects of blocking. Because the dialed digits control the switches directly, there must be a strict correspondence between the dialed digits, which represent the telephone number of the called party, and the physical contacts to which the called line is connected. The physical number is called the *office number,* and the listed number is called the *directory number.*

The two problems of direct progressive control are avoided with *register progressive control.* As shown in Figure 4.29, the line-finder switch connects the calling party to a register to receive and temporarily store all the dialed digits. After the called number has been completely dialed into the register, the register is connected

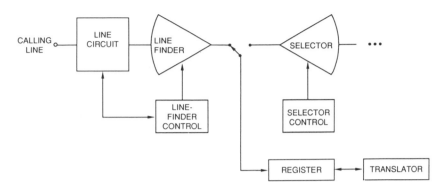

Figure 4.29 With register progressive control, the dialed digits representing the directory telephone number are entered and stored temporarily in a register. The register-stored digits can then be translated to the office number, which represents the actual location of the called party's contacts in the switching system. If blocking occurs, the register-stored digits can be reused for a second attempt.

to the selector and, in effect, "redials" the number. If blocking occurs, the number is still available in the register so that another redialing attempt can be made. A *translator* can be used with the register to translate the directory number into the office number needed to control the switches to reach the appropriate contacts. This means that the actual physical placement of a telephone line on the switch contacts does not need to be changed if a telephone number is changed and *vice versa.*

THE CROSSBAR SWITCHING SYSTEM

History and Invention

The Bell System's answer to the Strowger system was the *panel switching system,* which was first installed in 1921. The basic switch in the panel system was a horrendous affair able to access any 1 of 500 contacts, compared to Strowger's 100. This was accomplished by rotating cork rollers that moved the selecting contacts horizontally and vertically, much like the motion of a human operator at a switchboard. The basic switch created considerable electrical noise and needed much maintenance. Panel was a failure in its own day.

The real innovation in electromechanical switching was the Bell System's *crossbar* switching system, first installed in 1938 as the No. 1 crossbar system for large metropolitan offices. The No. 5 crossbar system, intended for rural and suburban offices, was first installed in 1948. A rural office crossbar system, the No. 3 crossbar system, was first installed in 1974. Crossbar switching has had a long life and was still in use in the early 1990s.

The Crossbar Switch

The basic crossbar switch is a matrix, or coordinate, switch with 10 horizontal rows and 20 vertical columns, as shown in Figure 4.30. Input lines appear on the 20 verticals, and the output connections are made on the 10 horizontals. Any vertical can be connected to any horizontal, and 10 simultaneous connections can be maintained by the switch. The actual electrical connections between verticals and horizontals are made by electromechanical contacts at the crosspoints that open and close.

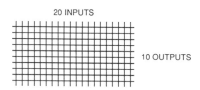

20 INPUTS

10 OUTPUTS

Figure 4.30 The basic switch used in the crossbar switching system is a matrix switch with 20 inputs along the verticals and 10 outputs along the horizontals.

The basic operation of the switch is ingeniously simple, yet difficult to visualize because the essential movements are very small and are hidden within the switch itself. As shown in Figure 4.31, there are five horizontal selecting bars that can each be rotated, or tipped, upward or downward through a small angle. These selecting bars are operated by electromagnets that receive their signals from the control section of the switching system. The verticals are addressed by 20 vertical hold bars, likewise operated by electromagnets that receive their signals from the control section. Operating the appropriate selecting bar either upward or downward and engaging the appropriate hold bar causes an electrical cross-connection to be made at the corresponding vertical and horizontal position.

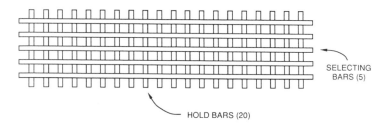

SELECTING
BARS (5)

HOLD BARS (20)

Figure 4.31 Five horizontal selecting bars each can rotate upwards or downward through a small angle in the crossbar switch. Twenty vertical hold bars can each be engaged to make a specific connection at the desired cross-connect.

Twenty short pins of flexible wire, called *fingers,* protrude from the back of each selecting bar (see Figure 4.32). These fingers protrude backward into the switch assembly. As the selecting bar is tipped upward or downward, these fingers likewise move up or down. The fingers thus have three possible positions: at rest, up, or down. The hold bars rotate and press on the fingers causing them to move sideways.

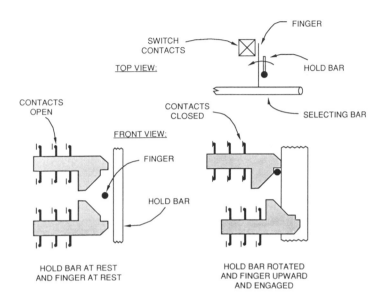

Figure 4.32 Twenty small, flexible pins, or *fingers* as they are called, protrude along the back of each selecting bar. As the bar rotates, these fingers move upward or downward. When the chosen hold bar rotates, it clamps onto the appropriate single finger and locks it in place in the switch contact assembly. The selecting bar can then be released and the hold bar still holds the selected finger. The engaged finger slips into a notch in a piece of phenolic material. The sideways movement of the finger causes the phenolic to move and this causes the switch contacts attached to the phenolic to be closed.

The switch contacts at each crosspoint are attached to a piece of phenolic material that has a small notch at the end close to the finger. The contacts are in pairs. The phenolic is notched so that if the finger is in the rest (or unselected) position, the hold bar will push the finger sideways between both phenolic pieces without touching either of them. However, if the finger is moved up or down by rotation of the selecting bar, the finger will engage the notch and will be held there by the action of the hold bar. This causes the switch contacts to close. The selecting bar can be released, but the hold bar will lock the flexible finger in position, thereby maintaining the connection. Other connections can then be made with the same selecting bar by activating other hold bars.

The contacts in a crossbar switch need only to close and open, which requires very little physical movement. Hence, wear and associated noise are at a minimum. Since more than one connection can be made by each crossbar switch, fewer switches are needed to make a switching network. The very simplicity of the crossbar switch minimizes costly maintenance. These advantages coupled with the use of a sophisticated control mechanism made the crossbar switching system a true winner. Finally, the Bell System had its answer to the Strowger system.

Crossbar System

The crossbar switching system uses a centrally located control mechanism that is shared in common by all the switches in the switching network. This type of control, therefore, is called *common control*. The actual control is performed by electromechanical relays called *markers*. The markers perform translation, test switches to determine whether they are in use, test different paths through the switching network, and issue the signals necessary to operate the switches. In effect, the markers are permanently programmed (called hard-wired) computers based on relay technology. The common control markers are used only briefly on each call. Once used, the markers drop out, and the switching network maintains the voice connection for the duration of the call.

As shown in Figure 4.33, the switching network is organized into two distinct groups of multistage crossbar switching matrices. The *line links* connect to the subscriber lines, and the *trunk links* connect to intraoffice and interoffice trunks. The line links and trunk links are situated in *frames* (see Figure 4.34), and the two frames are connected together by wires called *junctors*. The junctors are situated together in their own frame.

The markers that control the line links and the trunk links access them through *connectors*. The originating register supplies dial tone and stores the dialed telephone number. Telephone numbers are translated into locations on the line link frame by the *number group*. An intraoffice call is connected back to lines within the office by intraoffice trunks at the trunk link frame. Interoffice trunks are controlled by the markers through connectors and various registers.

ELECTRONIC SWITCHING SYSTEMS

AT&T No. 1 ESS® Switching System

The markers in a crossbar system are hard-wired computers, and their nonprogrammable inflexibility is a serious disadvantage. The idea of using a digital com-

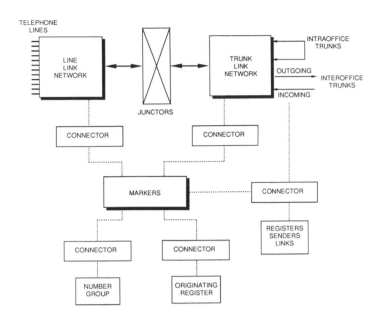

Figure 4.33 A major innovation with the crossbar switching system is the use of sophisticated control that is shared in common by all the switches and calls.

puter, with its inherent programmable flexibility, to control the operation of the switching network was a natural progression in the evolution of switching technology. Digital computers are electronic devices, unlike the electromechanical relays used in markers in the crossbar system, and hence the new generation of switching technology was called an *electronic switching system* (ESS).

The first electronic switching system developed by AT&T was the No. 1 ESS® system. It was first installed in 1965 and was designed to serve from 10,000 to 65,000 lines at a maximum of 25,000 calls in the busy hour. The No. 1A ESS system was introduced in 1976 and served as many as 128,000 lines at a maximum of 240,000 calls in the busy hour.

Reed Switch

The speed and flexibility of a programmable computer to control the switching network necessitated a new basic switch. The new switch was a *reed switch,* consisting of two reed contacts sealed in a small glass envelope, as shown in Figure 4.35. The application of an external magnetic field causes the contacts to close and open. The reed switches are organized into basic eight-by-eight coordinate arrays, or switching matrices.

Figure 4.34 Photograph of a crossbar switching system. (Courtesy of AT&T Bell Labs.)

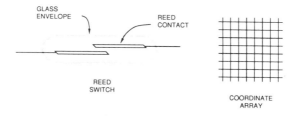

GLASS
ENVELOPE

REED
CONTACT

REED
SWITCH

COORDINATE
ARRAY

Figure 4.35 A simple reed switch is used in the AT&T No. 1 ESS switching system. The reed switches are organized into basic eight-by-eight coordinate arrays, or switching matrices.

Two approaches have been used to operate the reed switches. The initial approach uses two plates of a special two-state magnetic alloy, called *Remendur.* The plates are mounted on both sides of the reed switch. The special alloy has the property that when momentarily magnetized in either polarity by the application of an electric magnetic field, the alloy retains the magnetism and polarity. In effect, the alloy is a magnetic memory. In one polarity, the alloy causes the reed contacts to close; in the opposite polarity, the alloy causes the reed contacts to open. Short pulses of electric current change the magnetism of the alloy. The assembly of reed switch and Remendur alloy is called a *ferreed switch.*

A later approach used the Remendur alloy for the reeds themselves. Thus, the magnetic polarity of the reeds could be reversed by the application of external electric currents, and the reeds would retain the polarity. In one polarity, the contacts are open; in the opposite polarity, the contacts are closed. This type of switch is called a *remreed switch.* It is smaller and more reliable than the ferreed switch that it replaced.

Stored Program Control

The use of a programmable, digital computer for control in the No. 1 ESS system is called *stored program control.* Dual processors continually check each other for faults. The programs to control the processors and translation information are stored in a semipermanent memory, called the *program store.* The status of lines and trunks, called numbers, and other temporary information are stored in the *call store.* As shown in Figure 4.36, the central control periodically scans the status of lines and trunks over *scanners.* Signals to operate the switches are sent over *distributors.*

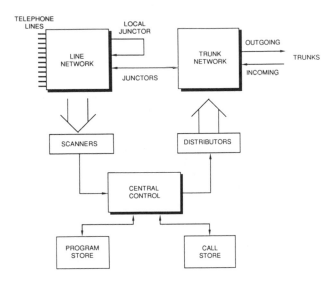

Figure 4.36 Dual processors form central control in the No. 1 ESS system. Access to and from the switches in the switching network is over scanners and distributors.

The No. 1 ESS system has sold well, and roughly 1000 machines are in service. All of its features are controlled by software, and updating that software has become a big chore, primarily because of the sheer size of the programs. Many of the new, functional services available today from local telephone companies are obtained over No. 1 ESS machines.

DIGITAL SWITCHING SYSTEMS

AT&T No. 4 ESS® Switching System

The No. 4 ESS® digital switching system, shown in Figure 4.37, was first installed in the Bell System in 1976. The No. 4 ESS system is used in toll and tandem switching, serving a maximum of 53,760 two-way voice circuits and 550,000 call attempts per peak hour. The switching network combines time and space switching in a

Figure 4.37 Photograph of AT&T's No. 4 ESS® switching system. (Photo courtesy of AT&T Network Systems.)

TSSSST order. The technology used in the No. 4 ESS system is all solid state, using integrated circuits where possible.

Time switching in the No. 4 ESS system is accomplished in time-slot interchange units composed of many buffer memories (see Figure 4.38). Each buffer memory can store 128 time slots. The 128 slots are used for 120 digital voice circuits, and the remaining 8 slots are used for maintenance purposes. Eight buffer memories, including one buffer memory as a maintenance unit, comprise the input to a TSIU, as shown in Figure 4.39. These eight buffer memories are assigned uniformly to eight output buffer memories for load-spreading purposes. The assignment of the specific slots in specific buffer memories is performed by a device called a *decorrelator*. Each output line, called an A-link, from the output buffer memories

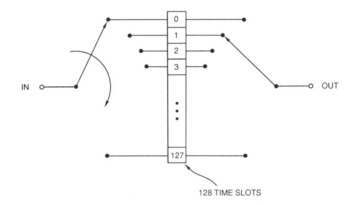

Figure 4.38 The buffer memories used in the AT&T No. 4 ESS system contain 128 time slots with each slot containing 8 bits. The time slots are written into the buffer in sequence, but are read out in a different order determined by the desired connections to be made.

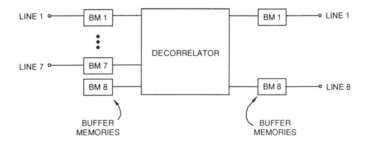

Figure 4.39 A time-slot interchange unit in the No. 4 ESS system consists of eight input buffer memories and eight output buffer memories. A decorrelator spreads the load uniformily across all the buffer memories. The TSIU serves seven digital lines each carrying 120 digital circuits.

carries 7/8 of the input circuits, or (7/8)120 = 105 circuits. Each output line has 128 time slots per 1/8000 second. A single TSIU can serve a total of 7 × 120 = 840 digital circuits. The input and output lines to the TSIU are coaxials.

A total of 128 TSIUs is connected at the input to a *time-multiplexed coordinate switch,* and another 128 TSIUs are connected to the output of the coordinate switch, as shown in Figure 4.40. Because each TSIU has eight output lines, the time-multiplexed switch must have 128 × 8 = 1024 inputs and another 1024 outputs. The time-multiplexed switch is a huge, time-multiplexed space switch in the form of a 1024-by-1024 matrix switch. Each digital circuit switched through the switching network is one way, and hence two such one-way circuits are needed to create a two-way circuit. Therefore, the switching network in the No. 4 ESS system can handle a maximum of 1024 × 105 = 107,520 one-way digital trunk circuits, or 53,760 two-way digital trunk circuits.

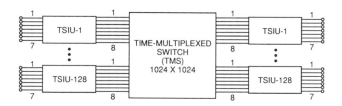

Figure 4.40 The overall architecture of the switching network used in the No. 4 ESS system is a basic time-space-time digital switching system. The space switch is a time-multiplexed solid-state coordinate switch with as many as 1024 inputs and 1024 outputs. Each cluster of 8 inputs or 8 outputs is served by a TSIU.

The time-multiplexed switch (TMS) performs its space switching in four stages to distribute the load uniformly, as shown in Figure 4.41. The eight output lines from each TSIU are connected to an 8-by-8 time-multiplexed space switch for uniform load distribution. The outputs from this load-spreading switch are connected in pairs to four time-multiplexed space switches, which are each capable of handling 256 inputs. The 256 outputs from each of these switches are connected to another time-muliplexed space switch, and again there are four of these switches. Another 8-by-8 time-multiplexed space switch interfaces to each output TSIU. In this manner, the switching is accomplished in four stages of space switching with one stage of time switching at the input and output, respectively. This then is a TSSSST switching network.

The 256-by-256 time-multiplexed space switches are broken down into two stages of switching performed by 16-by-16 switches interconnected by intraswitch lines, called B-links, to distribute the load, as shown in Figure 4.42.

Conventional analog circuits need an interface unit to convert them to the digital format required by the No. 4 ESS system. This conversion is performed by

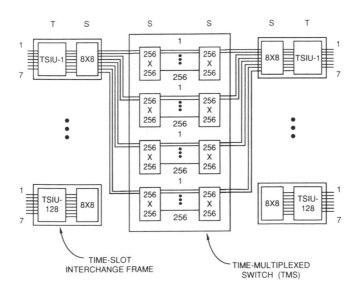

Figure 4.41 Switching in the No. 4 ESS system is accomplished in a TSSSST order. A time-slot interchange unit and an 8-by-8 time-multiplexed space switch for a time-slot interchange frame.

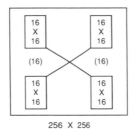

Figure 4.42 Each 256-by-256 switch is comprised of two stages of 16-by-16 switches. B-links interconnect the two stages.

a *voiceband interface unit* (VIU), which digitizes and multiplexes 120 voiceband circuits at 64,000 bps each. Up to six paths through the switching network of the No. 4 ESS machine can be synchronized to set up a 384-kbps digital connection.

Northern Telecom DMS® Switching System

Northern Telecom manufactures the DMS™ (for *D*igital *M*ultiplexed *S*ystem) series of digital switching systems. A photograph of a DMS machine is shown in Figure 4.43. The DMS-100 family of systems is intended for local offices, and the

Figure 4.43 Photograph of Northern Telecom's DMS® switching system. (Photo courtesy of Northern Telecom.)

DMS-200 family is intended for toll and tandem applications. The various machines are all similar in basic structure; the differences are in software and in peripheral modules. The DMS-200 first saw service in 1978 and could handle 61,444 trunks.

The overall architecture of the DMS system, shown in Figure 4.44, is fairly simple and consists of two independent, fully duplicated digital switching networks organized into two planes. In case of trouble, each switching network can take over for the other and can carry the full load. The switching networks interface with lines and trunks through various traffic interface modules. Control is by two independent central processing units (CPU) with one CPU active and the other standby. The two CPUs operate in synchrony and continuously check each other, although one is in control. Each CPU has its own program store and data store. The actual programs used to control the operation of the system are contained in program store, which contains as many as eight million 17-bit words of memory. Translation information, temporary call information, and other administrative information are stored in data store. The CPUs communicate with the switching networks through two *central message controllers* (CMC). Magnetic tape units, visual display units, and teletypewriters are connected to the CMC through two input-output controllers

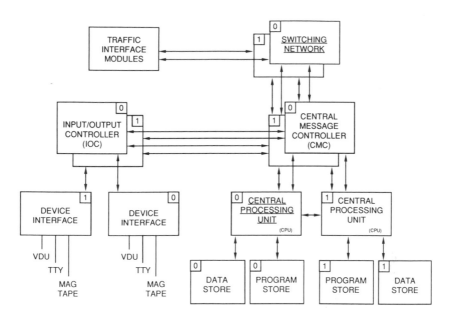

Figure 4.44 The overall architecture of the Northern Telecom DMS® digital switching system consists of a switching network organized into two planes, various traffic interface modules, and a fully duplicated common control. The duplicated CPUs continuously check each other, and in case of a fault, one takes over for the other. The control portions of newer DMS systems are interconnected by a more flexible bus structure.

(IOC). In newer DMS machines, the functional elements of the control section are interconnected by a bus structure operating at 128 Mbps. The bus interconnection is more flexible and facilitates multiple processors, memories, and switching architectures.

Virtually every unit in the DMS switching system is duplicated. This is both for load-sharing purposes and also for possible trouble so that one unit can take over for the other. Extremely reliable operation is a major consideration in switching system design, particularly for large toll switching systems that handle considerable amounts of network traffic.

The switching network is organized into two separate but fully duplicated planes, numbered 0 and 1. Each plane is modularized into 32 network modules (NM), numbered 0 through 31, as shown in Figure 4.45. Each network module has 64 two-way digital inputs and 64 two-way digital ouptuts. Junctors connect each module to each of the other modules (inter-NM junctors) and also to itself (intra-NM junctors). A junctor is a communication circuit that carries signals within the switch itself. A junctor is an intraswitch trunk.

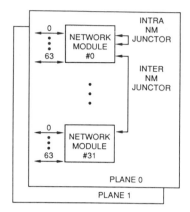

Figure 4.45 Each plane of the switching network of the DMS system consists of 32 network modules each capable of switching the time slots in 64 two-way lines. The lines between the network modules are the junctors. The two planes are fully replicated, and each plane can carry the full switching load.

The specific architecture for the switching network has evolved over time. In older DMS machines, the switching network was organized into 32 network modules that performed the switching in two stages of TST switching, as shown in Figure 4.46. Clusters of 8 digital input lines were served by a TST switch so that a total of 8 TST switches were needed to handle the 64 input lines. Another 8 TST switches handled the output lines in the second stage of switching through the network module. Each line carried 32 time slots. Later, the switching module was changed to an STTS architecture serving 64 input lines and 64 output lines.

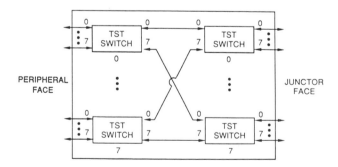

Figure 4.46 At one time, the switching module consisted of two stages of time-space-time (TST) switching. Seven TST switches served the peripheral face, and another seven TST switches served the junctor face.

The newest DMS machines use optical fiber for input and output from the switching network, called "E-NET" for *E*nhanced *Net*work. Each fiber carries 512 time slots. The switching network is a single-stage, nonblocking, time switch capable of switching 131,072 one-way digital circuits or 65,536 two-way digital circuits. The switching network, shown in Figure 4.47, consists of eight horizontal buses for input and eight vertical buses for output. Each bus carries 16,384 time slots in 1/8000 second. Various interfaces are used to combine signals for input to and output from the buses. The DS512 fiber interface handles four fiber links with each link carrying 512 time slots. Eight such fiber interfaces can be served by one bus. A TSIU is located at each of the 64 crosspoints between the vertical and horizontal buses. The input to the TSIU comes from a vertical bus, and the output goes to a horizontal bus. Each TSIU stores 16,384 time slots in a double-buffered configu-

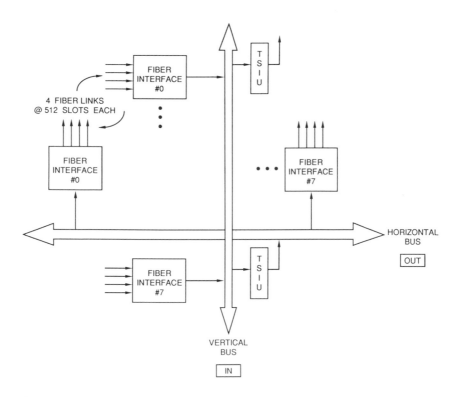

Figure 4.47 The architecture of the E-NET switching network can be visualized as a large, single-stage, nonblocking time switch capable of switching 131,072 one-way digital circuits. The actual switching is accomplished by 64 time slot interchange units located between vertical and horizontal buses. The vertical bus carries the input slots, and the horizontal bus carries the switched output slots. The DS512 fiber interface connects four fiber links to a bus.

ration so that the delay through the TSIU is always a fixed 125 μs for all slots. Northern Telecom calls the TSIU a "double-buffered 16K×16K time switch." Here, the "K" refers to the binary K of 1024 instead of the decimal "k" of 1000.

E-NET is a most impressive digital switching network. The processor that controls it has much to do to assure that each time slot makes its way to its destination at precisely the correct time and through the correct time-slot interchange unit. E-NET can be visualized as a single, very large time switch with multiple input links and multiple output links—an approach discussed in a previous section of this chapter.

Analog and digital lines and trunks interface to the switching system through traffic interface modules. The two-way digital output lines, called links, from each traffic interface module connect to both planes of the switching network. The outgoing link connects directly to both planes, while the incoming link has a time-multiplexed switch to choose the appropriate plane. In effect, each two-way link from a traffic interface module sends to both planes and receives from one plane, as shown in Figure 4.48. The three types of traffic interface modules are a *line concentrating module* (LCM), a *trunk module* (TM), and a *digital trunk controller* (DTC). Different modules are configured depending on the specific application of the DMS switching system.

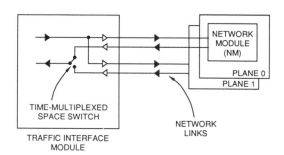

Figure 4.48 A time-multiplexed space switch in each traffic interface module switches between the two planes of the switching network.

A line concentrating module serves 640 analog local-loop circuits or lines. Each local loop is connected to a line card, shown in Figure 4.49, that supplies battery and functions (known collectively as BORSCHT). The line card is where analog-to-digital conversion takes place for each local-loop termination. The output from the line card is a two-way PCM, or digital, signal at 64 kbps. A total of 32 such PCM signals is time multiplexed together by the bus interface card, as shown in Figure 4.50. The bus interface card has 30 output time slots. The two-way digital circuit from the bus interface card is called a *terminal link* and operates at 2.56 Mbps.

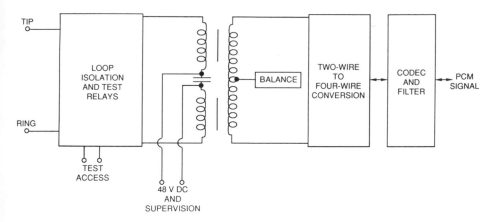

Figure 4.49 Each local-loop line is served by its own line card. The line card supplies battery, supervises the loop, gives access to various line tests, performs two-wire–to–four-wire conversion, and performs analog-to-digital conversion. The ouptut from the lines card is a pulse code modulation signal.

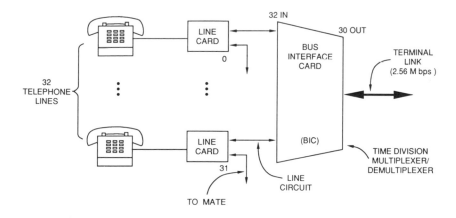

Figure 4.50 The bus interface card performs time-division multiplexing and demultiplexing for the telephone lines served by 32 line cards. The output from the bus interface card is called a *terminal link*. A terminal link carries a digital signal at 2.56 Mbps. A small amount of concentration occurs in the bus interface card since its 32 input voice-channel time slots are reduced to 30 output voice-channel time slots. Two additional slots are created in the terminal link signal for signaling and control purposes.

Each of the 32 time slots carried on the terminal link contains 10 bits. PCM voice (and possibly data) is carried in 30 of the time slots, which means that because there are 32 input circuits, a small amount of concentration takes place in the bus interface card. Each 10-bit time slot contains 8 bits of PCM voice and data information along with 2 bits for signaling and intra-switch communication purposes. For example, for a signal coming from the line card over a line circuit, one of the two bits indicates whether the line is on- or off-hook. The remaining two of the 32 time slots carry control information.

Twenty bus interface cards, numbered 0 to 19, connect to a time-multiplexed space switch, as shown in Figure 4.51. The multiplexed switch ultimately serves 640 local-loop circuits connected to their line cards and performs concentration of the loop circuits. The input to the time-multiplexed space switch is 20 terminal

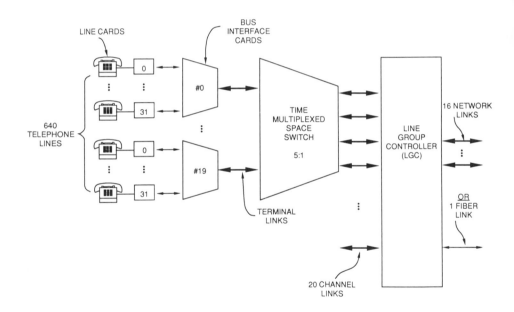

Figure 4.51 A total of 20 bus interface cards is served by a time-multiplexed space switch. For the 5:1 concentration shown here, the 600 input voice-channel time slots are reduced to 120 voice-channel time slots. These 120 time slots are carried over 4 two-way channel links each operating at 2.56 Mbps. An additional 2 time slots for signaling and control purposes have been added to each channel link, so that each link actually carries 32 ten-bit slots. As many as 20 channel links connect to the line group controller, which performs call-processing functions, switching between the two planes of the switching network, and a small amount of concentration. In newer DMS systems using E-NET, a single two-way fiber link (two fibers) connects the LGC to the switching network; in older DMS systems, as many as 16 network links, each operating at 2.56 Mbps over copper wires, connect the LGC to the switching network.

links, with each terminal link carrying 30 two-way digital voice and data circuits. The output from the time-multiplexed switch is from 2 to 6 channel links, depending on the concentration ratio, with each channel link likewise carrying 30 digital voice and data circuits. For a concentration ratio of 5:1, the 600 input circuits are concentrated onto 120 output circuits carried on 4 channel links. Each channel link operates at 2.56 Mbps in each direction. As many as 20 channel links connect to a line group controller (LGC), which performs call-processing functions and switching between the two network planes. From 2 to 16 network links, each operating at 2.56 Mbps over copper wire, connect the line group controller to the switching network. In newer DMS systems using E-NET, the information in up to twenty 2.56-Mbps channel links is combined by the line group controller and transmitted over a single optical fiber link carrying 512 time slots (a so-called DS512 signal).

Two line modules operate together as mates. The processor in each line module can access all the cards in the other line module. This is so that in case of trouble with one processor, the other processor can take over, even though the traffic-handling capacity would be reduced. A master processor is responsible for controlling the operation of the line module, and a signaling processor is responsible for controlling line scanning, ringing, and other supervisory and signaling functions.

The DMS system can be used for a wide variety of switching applications, including local, toll, tandem, cellular radio, and interexchange. It interfaces with various loop-carrier systems, such as AT&T's SLC® 96 system and Northern Telecom's DMS-1 urban loop carrier.

AT&T 5ESS® Switching System

The AT&T 5ESS® digital switching system, shown in Figure 4.52, is intended for a wide variety of applications, such as local, toll, tandem, and PBX switching. The first system was cut over in March 1982. The 5ESS system fully configured can handle 100,000 lines and 650,000 conventional telephone calls per hour. As is usual with digital switching machines in general, the technology continues to progress, and newer machines are made with even greater capacities. For example, the 1983 version of the 5ESS system had only 30 switching modules and could handle 50,000 lines and 130,000 calls per hour.

The overall architecture of the 5ESS system, shown in Figure 4.53, is fairly simple and consists of a number of switching modules (SM) and a single, central communication module (CM). The switching modules are time-slot interchange units that perform time switching. The communication module is a time-multiplexed space switch that performs space switching between the switching modules. An administrative module (AM) is responsible for such functions as billing, call translation, trunk routing, and other administrative features. Control is distributed throughout the switching system, and each switching module has its own processor

Figure 4.52 Photograph of AT&T's 5ESS® switching system. (Photo courtesy of AT&T Network Systems.)

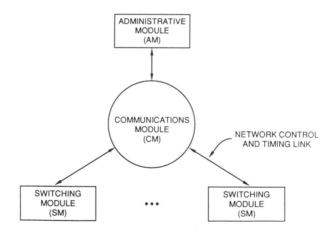

Figure 4.53 The overall architecture of the AT&T 5ESS digital switching system consists of many switching modules that interface to analog and digital lines and trunks and perform time-slot interchange. Switching between modules is performed by a time-multiplexed space switch in the communication module. Control is distributed throughout the system, and only administrative functions are performed centrally by the administrative module.

and memory for module control. The switching modules communicate with the communication module over two-way fiber links with each multimode fiber in a pair carrying 32.768 Mbps. Switching modules can be located remotely: a remote switching module (RSM) up to 100 miles, and an optically remote module (ORM) also up to 100 miles using a fiber connection back to the communication module. The administrative module communicates with the system over two 16-bit links connected to the communication module.

As shown in Figure 4.54, the switching module is primarily a time-slot interchange unit with various interface units connecting to analog and digtial lines and trunks. The switching module communicates with the time-muliplexed switch in the communication module over network control and timing links. These links are two pairs of multimode fiber. Each fiber carries 256 time slots in 1/8000 second: 255 time slots carry digital voice and data, and the remaining slot carries control messages. Each time slot consists of 16 bits: 8 bits for digital voice and data with the remaining 8 bits for signaling and control, such as on- or off-hook indication, the status of the slot, and parity.

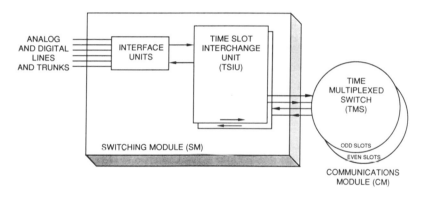

Figure 4.54 The time-multiplexed space switch in the communication module is organized into two planes: one for odd-numbered time slots and the other for even-numbered time slots. The switching module consists of time-slot interchange units: one for slots being sent to the TMS and the other for slots coming from the TMS. Various interface units connect to analog and digital lines and trunks.

The time-multiplexed switch is organized into two planes: one plane serves 256 odd-numbered time slots, the other serves 256 even-numbered time slots. The TMS is a high-speed matrix switch with a maximum of 192 ports capable of serving 190 switching modules. The information in a time slot passing through a time-slot interchange unit can be delayed a maximum of 125 μs. The TMS, however, has zero delay. Each plane of the TMS must reconfigure its switching matrix 256 times every 1/8000 second, which works out to 2,048,000 reconfigurings per second. The communication module contains the network clock that synchronizes the total operation of the switching system.

As shown in Figure 4.55, a module control unit supervises the operation of the switching module; each SM has its own control unit, and thus control is distributed throughout the system. The optical signals carried over the fiber links are converted to electrical signals by dual-link interfaces (DLIs) in the switching module and the communication module. The various interface units in the switching module are connected to the TSIU over peripheral interface data buses (PIDB) carrying 510 two-way time slots. Each PIDB consists of serial data circuits with each individual circuit carrying 32 16-bit time slots. The output from the TSIU has 512 time slots although the input from the PIDB has 510 slots. The two additional slots are used for control information by the processor that controls the SM.

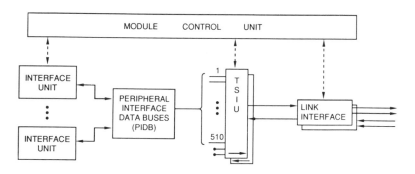

Figure 4.55 Each switching module has its own module control unit. The signals sent over the optical links between the switching module and the communication module are converted to electrical signals by dual-link interfaces located at each end of the fiber links. Peripheral interface data buses carry signals between the interface units and the TSIU.

The subscriber line unit (LU), shown in Figure 4.56 interfaces with local-loop telephone lines. A single LU serves from 256 to 640 lines, depending on the desired concentration for specific traffic. The lines are concentrated and accessed by a space-division matrix, or crosspoint, switch. This matrix switch is composed of solid-state gated-diode switches. Hybrid integrated circuitry (HIC) is used to package the gated diodes because it can withstand the high voltages of ringing and battery. The concentrator has 64 output lines, and thus 640 telephone lines can be served with a 10:1 concentration; 512 lines with 8:1 concentration; 384 with 6:1 concentration; and 256 lines with 4:1 concentration.

Each two-wire output line from the concentrator is connected to a channel circuit that supplies battery, performs two-wire–to–four-wire conversion, supervises the status of the loop, and performs analog-to-digital conversion. Thirty-two

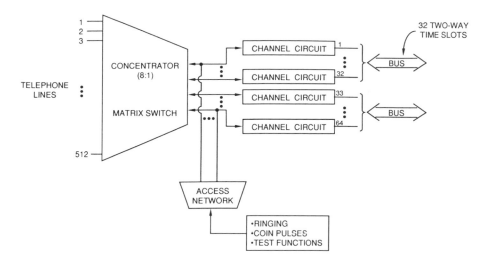

Figure 4.56 The subscriber line unit can serve from 256 to 640 telephone lines depending on the acceptable amount of concentration. An individual line is accessed by the concentrator, which is a space-division matrix switch using solid-state gated diodes at the crosspoints. Channel circuits supply battery, perform two-wire–to–four-wire conversion, supervise the loop, and perform analog-to-digital conversion.

channel circuits are multiplexed together onto a single two-way serial bus in the peripheral interface data bus. The two-wire lines from the concentrator can be accessed to supply ringing and coin control pulses and for network test functions.

The digital line trunk unit (DLTU), shown in Figure 4.57 interfaces with DS1 digital trunks at 1.544 Mbps, usually T1 facilities. The digital facility interface (DFI) handles a single DS1 trunk and converts the bipolar pulses of the DS1 to the unipolar pulses used in the 5ESS system. The 24 time slots in the DS1 are distributed over 32 time slots that are sent over the peripheral interface data bus: 24 slots carry voice and data information and 8 slots are evenly distributed as empty fill. The DLTU can have as many as 10 DFIs. In Europe, a digital trunk operates at a rate of 2.048 Mbps and contains 32 time slots; in this case, the DLTU can have only up to 8 such DFIs.

The trunk unit (TU) interfaces with 64 analog trunks and converts their analog signals into 64 time slots. The digital carrier line unit (DCLU) handles subscriber loop carrier systems, such as the SLC®96 system. A maximum of six SLC96 systems can be served. The DCLU has up to a 9:1 digital concentration so that if the maximum number of six SLC96 systems is served, the output of the DCLU is 64 time slots.

The integrated services line unit (ISLU) supports both analog lines and the digital subscriber line (DSL) interface that is used for ISDN applications. The ana-

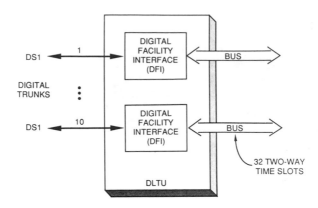

Figure 4.57 The digital line trunk unit (DLTU) connects DS1 digital to the data buses through digital facility interfaces (DFIs).

log lines are converted to digital signals. The ISLU uses digital concentration, as opposed to the analog concentration used in the subscriber line unit.

BORSCHT

A local exchange switching system that serves local-loop lines is responsible for performing a number of functions that are lumped together under the acronym *BORSCHT.*

The local switching office must supply battery feed at 48 V dc to power the telephone instruments connected to the local loops. The switching equipment must be protected from high voltages that might occur during a lightning strike on the line, and hence some form of overvoltage protection must be supplied. The local office supplies the ringing signal necessary to cause the telephone instrument to ring. The local switching machine needs to supervise whether the local line is in use or not, and hence supervision must be supplied. In the case of a digital switching machine, the analog signal on the local loop needs to be converted to a digital signal and visa versa, and hence coding and decoding must be supplied. The two-wire local loop needs to be converted to the four-wire signal required for modern switching systems, and hence a hybrid function must be performed. Last, various types of loop testing are needed.

Putting all this together, we have:

*B*attery feed,

*O*vervoltage protection,

*R*inging,

*S*upervision,

*C*oding and decoding,

*H*ybrid, and

*T*esting.

PRIVATE BRANCH EXCHANGES

Many organizations have a need for their own exclusive switching systems that are, in effect, a telephone switching exchange acting as a branch of a central office for the private use of the organization. Such switching systems are called *private branch exchanges,* or PBXs for short. The PBX usually is placed on the premises of the organization.

The various technologies used for PBXs closely parallel the development of the switching technologies and systems used for telephone network switching. The earliest PBXs were manual switchboards operated by human attendants, usally situated near the entrance to the facility so the operator could also serve as the receptionist. The attendant knew the comings and goings of all the staff and could take messages and forward calls as needed. Cordless switchboards were developed to ease the work of the attendant.

The role of the attendant was automated through the use of such electromechanical technologies as step-by-step and crossbar, adapted for use in PBXs. Such automated PBXs are called *private automatic branch exchanges* (PABXs). Time-division switching utilizing digital technology and computer control has created a whole new generation of PBXs, called *computerized branch exchanges* (CBXs). The term PBX is still used to refer generically to any type of private branch exchange, whether automatic, electronic, or computerized.

Some of the newest PBXs utilize time-division switching of pulse-code modulated (digital) speech signals. Special telephone instruments that perform analog-to-digital conversion are used and transmit digital signals over the local on-premises wiring. The digital signal can also represent text, and so these types of all-digital PBXs are particularly applicable to electronic mail and remote access to computerized databases. Since the speech signals have already been digitized, such PBXs frequently use T1 trunks for access to the local central office and the public switched network.

Modern PBXs can be quite large, serving thousands of telephone extensions at a single location. Many large companies have locations scattered about the country, and therefore it is quite natural to connect all the PBXs serving the locations through a private long-distance network. The result is a corporate communication network that can be quite complex and extensive. Synthesizing many different PBX vendors and network providers to create an integrated telecommunication facility is the job of today's corporate telecommunication manager.

Most PBXs allow individual telephones to dial outside directly over the public switched network, a feature that is called *direct outward dialing* (DOD). However, if the local telephone company assigns only a single telephone number to the PBX, then all incoming calls must be answered by an attendant who makes the final connection to the desired extension. An alternative approach is the use of recorded speech asking the calling party to touch-tone enter the extension number so that the PBX can automatically make the final connection. An even better solution is for the local telephone company to assign a whole exchange number to the PBX so that all incoming calls can reach any known extension directly simply by dialing the complete seven-digit number. This feature is known as *direct inward dialing* (DID).

PBXs have their advantages. Equipment is not rented from the local telephone company. The PBX users are in complete control of their telecommunication system and can do as they wish, and when they wish, without waiting for the telephone company to be available. However, particularly for a large PBX installation, the telecommunication manager in reality is operating a small telephone company with all the inherent problems. Another solution is the use of the switching system at the local central office as if it were a private branch exchange. This type of service, provided by the local Bell telephone company, is called *Centrex* service. It is offered on a leased-service basis and competes with the purchase of an on-premises PBX. Centrex service operates with standard telephones, and instructions are indicated by momentarily operating the hook, called *hook flashing,* which many users find confusing and disconcerting.

Telemarketing involves the use of the telephone, and usually a toll-free 800 number, for placing orders for goods and services. An agent answers the call and types the order into a computer. Many agents are available to answer the calls and a means is necessary to distribute the calls uniformly to available agents. This function is called *automatic call distribution* (ACD) and is performed by an automatic call distributor installed at the PBX or at the local office.

If one has only a small number of extensions, a simple *key telephone system* (KTS) will usually suffice. Keys on each telephone are depressed to seize a line. Typical key telephones have six keys, offering access to five lines with the sixth key a "hold" key. The keys are lighted and flash in different patterns to indicate whether the particular line is in use, on hold, or ringing. Older key systems utilizing electromechanical relays and multiwire-pair cables are being replaced by newer electronic systems that mimic the features of key telephone systems but work over only a few pairs of wires and utilize computer technology.

TRAFFIC

Switching and transmission systems are designed to handle the expected traffic. If a particular system is designed to be nonblocking, then all calls will be served. Most switching systems block some calls if they handle a large volume of traffic.

Some useful measures of traffic include the total number of calls served during a specified time period, usually the busiest hour for the system, called the *busy hour.* Usage can be specified in terms of the percentage of time a trunk or switch is in use. Overflow is measured as the percentage of calls that find some particular piece of equipment or the whole system busy.

Telephone traffic is measured in units of *hundred call seconds,* or CCS. The product of the number of calls multiplied by their average duration gives the traffic in call seconds. Dividing this number by 100 gives the traffic in hundred call seconds. For example, 20 calls each with an average duration of 120 seconds would generate $(20 \times 120)/100 = 24$ CCS. The *traffic density* is the number of CCS per time period, usually the busy hour. For example, a PBX might be specified as being able to handle a traffic density of 1500 CCS per busy hour.

Another measure of telephone traffic is the *erlang,* named after the Danish engineer and mathematician, A.K. Erlang. One erlang is equivalent to one call lasting one full hour. Because an hour has 3600 seconds, one erlang is the same as 36 CCS. If there were 100 percent occupancy of a telecommunication path or device, then the traffic carried would be specified as either 1 erlang or 36 CCS. The PBX with a busy-hour capacity of 1500 CCS could equivalently be specified as having a busy-hour capacity of $1500/36 = 42$ erlangs.

Traffic density is usually associated with a specific *grade of service.* The grade of service is the probability that a given amount of traffic will encounter blocking. For example, a grade of service of 0.01 for a traffic density of 1500 CCS means that, at a traffic density of 1500 CCS, one out of every 100 calls will be blocked.

A useful measure is the *load factor,* which specifies the average number of simultaneous calls during a specific hourly period, usually the busy hour. The load factor, C, can be calculated as the product of the number of calls originated, n, multiplied by the average holding time in seconds, h, divided by 3600, or

$$C = nh/3600$$

If the traffic is given in CCS, then the load factor is the traffic divided by 36. Consider a PBX that can carry 1500 CCS during the busy hour with a 0.01 grade of service. The load factor is $1500/36 = 42$. This means that the PBX can carry 42 simultaneous calls in the busy hour with a probability of blocking of 0.01.

Clearly, telecommunication systems can become congested. Models based on probability theory have been developed to characterize such congestion. The models relate the probability of blocking to the volume of traffic for different numbers of switching links, trunks, or *servers,* as they are called.

Two popular models are the *Poisson model* and the *Erlang-B model.* The Poisson model assumes that all blocked calls are retried within a relatively short period. This model is appropriate for describing the call attempts by a large number of customers, such as those served by a central office switching system. The Erlang-B model assumes that blocked calls are abandoned; it is appropriate for describing interoffice trunks. Tables are available that tabulate the capacity for different grades of service and different numbers of servers.

An example might help clarify the use of these models. Consider a PBX for which the total traffic during the busy hour has been measured to be 50 calls with an average duration of 5 minutes per call. The total busy-hour traffic is (50 × 5 × 60)/100 = 150 CCS. The acceptable grade of service is 0.01, which means that 1 call out of 100 will be blocked. The traffic and grade of service are looked up in the Poisson table, which shows that 10 servers are sufficient to handle the traffic. This means that a PBX capable of handling 10 simultaneous calls is sufficient, according to the Poisson model.

SWITCHING—AN ASSESSMENT

Indeed, the world of telecommunication is going digital. We saw in a previous chapter how optical fiber and time-division multiplexing are driving *transmission* toward digital. In this chapter, we saw how time-division switching and computer control are driving *switching* toward digital.

Voice communication is an analog affair. Hence, there will always be a need for analog-to-digital conversion. Most telephone instruments and, of course, the local loop are analog. Every local loop travels all the way back to the central office, thereby contributing to all the wire found there. Optical fiber can multiplex together hundreds, or more, of telephone conversations and can eliminate the need for a physical wire pair for each conversation. Some central office switching systems have switching modules that can operate remotely from the central office. Coupling these remote modules with fiber creates a new local network topology in which only the last few hundred feet to each telephone is wire pair. The wire pairs are brought to a remote switching module for switching and multiplexing, as shown in Figure 4.58. If this trend were to continue, the role of the central office as a local switching point would surely change, and the central office would offer primarily access to interexchange carriage.

The newer digital switching systems use optical fiber to carry signals between interface modules and the switching network. The conversion of light signals to electrical signals would be an unnecessary chore if light signals could be directly switched on fiber. So-called *photonic switching* is an area of active research and might indeed find application in future switching machines.

Many telephone companies invested heavily in replacing step-by-step and crossbar systems with AT&T's No. 1 ESS system. The No. 1 ESS system is a computer-controlled *space-division* switching system. The newer systems, such as the AT&T 5ESS system and the Northern Telecom DMS system, are computer-controlled *time-division* switching systems. Because most trunks are digital, these newer digital switching systems interface easily with digital trunks. They also are physically smaller than the No. 1 ESS machine. Thus considerable pressure exists

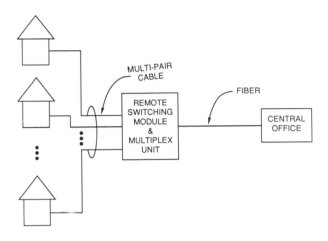

Figure 4.58 Remote switching modules and multiplex units are changing the topology of the local network.

to replace No. 1 ESS machines with the newer all-digital machines, even though the older No. 1 machines are still functioning perfectly and have years remaining on their useful service lives.

Electromechanical switching automated work previously performed by human operators and allowed telephone service to be universally affordable by everyone. However, electromechanical technology was inflexible and did little more than switch telephone conversations. The functionality provided by human operators was lost. No longer could calls be forwarded, messages be taken, and calls interrupted. Now, with computer control and the "intelligence" that it provides, flexibility and functionality have returned again to telephone service. Such services as call forwarding, call interrupt, voice mail, and calling-number identification are again available, but at a price easily affordable to everyone.

REFERENCES

Gurrie, Michael L., and Patrick J. O'Connor, *Voice/Data Telecommunications Systems,* Englewood Cliffs, NJ: Prentice-Hall, Inc., 1986.

Hayward, W.S., Jr., ed., "The 5ESS Switching System," *AT&T Tech. J.* (special issue), Vol. 64, No. 6, Part 2, July–August 1985.

Hills, M.T., *Telecommunications Switching Principles,* Cambridge, MA: The MIT Press, 1979.

"No. 1 Electronic Switching System," *Bell Syst. Tech. J.,* Vol. XLIII, No. 5, Parts 1 and 2, September 1964.

"No. 4 ESS," *Bell Syst. Tech. J.,* Vol. 56, No. 7, September 1977.

Schindler, G.E., Jr., ed., *A History of Engineering and Science in the Bell System: Switching Technology (1925-1975),* Murray Hill, NJ: Bell Telephone Laboratories, Inc., 1982.

Talley, David, *Basic Telephone Switching Systems,* Rochelle Park, NJ: Hayden, 1979.

Terry, J.B., D.R. Young, and R.T. Matsunaga, "A Subscriber Line Interface for the DMS-100 Digital Switch," *IEEE National Telecommunications Conference (NTC) Digest,* Vol. 1, 1979, pp. 28.3.1–28.3.6.

Terry, J., H. Krausbar, and J. Hood, "DMS-200 Traffic Peripherals," *IEEE Communications Conference Record (ICC'78),* 1978, pp. 32.3.1–32.3.5.

Chapter 5
SIGNALING

INTRODUCTION—MANUAL EXCHANGE SIGNALING

Telephone service involves a vast network of transmission and switching facilities that is controlled by the telephone user. Signals of various kinds are sent over that network to control its operation and to indicate its status. *Signaling* deals with those control and status signals. One way to understand some of the functions of signaling is to consider manual switching by human operators.

In the early days of telephony, exchange service was accomplished with manual switching by a human operator. The calling subscriber first alerted the operator by turning the crank on the telephone, which caused a lamp to flash on the panel at the exchange office. The operator saw the flashing lamp and plugged in on that line. The calling subscriber then verbally gave the operator the name or number of the called party. The operator then visually checked the lamp associated with the called party's line to determine whether the line was available. If it were not, the operator verbally informed the calling subscriber that the called party's line was in use. If the line were available, the operator rang the called party's phone and made the final connection of the two parties. Lamps indicated when a party hung up, and the operator would then take down the connection.

If the call were from one exchange to another, the operator at the calling party's exchange used special interexchange lines, called trunks, to reach an operator at the called party's exchange. The number being called was passed verbally from operator to operator, and the operator at the called party's exchange made the final connection.

Some of the signaling functions that occurred during this manual connection were the alerting of the operator that service was desired, the verbal specification of the called party's identity or address, the supervision by lamps of the status of the telephone lines or circuits, and the reporting by the operator of the progress of the call to the calling party.

Clearly, the making of a telephone connection during the early days of telephone service involved a large amount of human labor. Over the years, technology

---

reduced and finally eliminated all human labor in making a telephone connection. This was accomplished through automated switching machines and various electrical signals to request service, forward telephone numbers, and make the actual connection of circuits.

FUNCTIONS

The four general functions of signaling in modern telephony are:

1. alerting,
2. transmitting address information,
3. supervision, and
4. transmitting information.

Alerting deals primarily with requests for service, facilities, and action of some kind. The initial request by the subscriber for service is an alerting function. The local office might send alerting signals to other offices requesting interoffice trunks. A local switching system alerts the called party to answer the telephone.

The telephone number, or *address,* of the called party must be transmitted from the calling party to the local switching machine. This is accomplished either by dial pulses or by tones. The switching machines in the network need to have the address so that the call can be completed. These are some examples of the *transmission of address information.*

Switching machines need to know whether lines and circuits are idle or in use. They also need to know when a seized circuit is no longer needed and can be released for reuse. The status of circuits and lines thus needs to be *supervised.*

Information signals, such as dial tone, busy tone, and various recorded announcements, must be transmitted to the calling party.

SUBSCRIBER-LOOP SIGNALING

Two major realms of signaling are *subscriber-loop signaling* and *interoffice signaling.* We first describe how the four functions of signaling are accomplished using the subscriber's local loop.

The alerting function of requesting service occurs when the handset is lifted from its cradle, or goes off-hook, thereby causing a direct current to flow through the loop. This flow of direct current is sensed by the switching equipment at the central office and is interpreted as a request for service. The flow of direct current continues as long as the telephone is in use, and by supervising this flow, the switching equipment senses that the line is still in use. When the flow of direct current ceases, the switching equipment knows that the conversation has ended. Thus, direct current, or dc, signaling on the subscriber loop is used for alerting and supervisory functions.

Address information can be sent over the subscriber loop in two ways. The flow of direct current can be interrupted by a switch in the telephone dial to generate dial pulses, which are counted by equipment in the switching machine at the local office. Dial pulses occur at a rate of about 10 pulses per second. The second way for transmitting address information over the subscriber loop is in the form of unique combinations of two pure tones, called *touch-tone dialing.*

Information is transmitted over the subscriber loop either as audible tones or as recorded announcements. Four major audible tones are dial tone, line-busy tone, ring-back tone, and trunk-busy tone. Four different frequencies (350, 440, 480, and 620 Hz) are used, either singly or in combination with each other. Distinctive timing patterns are also used.

Dial tone is a continuous signal formed by the simultaneous transmission of a 350-Hz tone and a 440-Hz tone. Such pure tones are also known as sine waves. Ring-back or audible ringing is the ringing signal heard by the calling party. The ring-back signal is formed by the simultaneous transmission of a 440-Hz tone and a 480-Hz tone. The combination is on for two seconds and off for four seconds and then repeats. The line-busy signal is formed by the simultaneous transmission of a 480-Hz tone and a 620-Hz tone. The combination is on for 0.5 second and off for 0.5 second and then repeats. The trunk-busy signal indicates that either switching equipment or transmission facilities are not available to complete the call, and the call should be retried. The trunk-busy signal is also known as the reorder signal. It is composed of the same tones as the line-busy signal but is repeated at twice the rate, namely, 0.25 second on and 0.25 second off.

An alerting signal that is sent over the subscriber loop but that is not audible is the ringing signal that causes the called telephone to ring. The ringing signal is a tone of 75 volts (rms) at a frequency of 20 Hz. The ringing signal is on for two seconds and off for four seconds and then repeats this pattern until the called telephone answers the call.

INTEROFFICE SIGNALING

The oldest and most basic form of signaling to indicate the availability of interoffice trunks is direct current, or dc, signaling. The presence or absence of dc on a trunk indicates whether the trunk is idle or in use. Normal and reverse polarity of the dc is sometimes also used for interoffice trunk signaling.

Long-distance circuits carry only alternating current (ac), and hence dc signaling cannot be used on such trunks. Thus, a single-frequency (SF) tone is used, either within the voiceband (200 to 3400 Hz) or outside the voiceband (3700 to 3825 Hz). One popular scheme is the use of a 2600-Hz single-frequency tone to indicate the availability of a trunk. The 2600-Hz tone is placed on idle trunks. Once a trunk is seized, the tone is turned off by the near office, and this signals the distant office that a call is about to be placed over that trunk. Address information is then sent over the trunk to the distant office, using multiple-frequency key pulsing (described below). When the called party answers, the distant office seizes a return trunk and turns off its 2600-Hz tone. The two-way voice connection is thus established, and billing and conversation begin.

The address information is sent over the seized trunk by using combinations of two tones sent at a rate of 10 combinations per second. The tones are at frequencies of 700, 900, 1100, 1300, 1500, and 1700 Hz. Using all the possible two-frequency combinations of these tones, it was possible to encode all ten digits and six control functions. This type of in-band signaling is called *multiple-frequency key pulsing* (MFKP).

Though quite simple, in-band signaling has problems. For one, some dishonest subscribers discovered that they could generate their own tones for the fraudulent purposes of avoiding toll charges. For another, some speech signals could cause accidental disconnections. Out-of-band signaling avoided these problems but reduced the usable bandwidth for the speech signal. All these problems are avoided with common channel interoffice signaling.

COMMON CHANNEL INTEROFFICE SIGNALING

In 1976, a new interoffice signaling scheme was first installed in the Bell System, and, though undergoing steady enhancement, is currently in use on virtually all interoffice trunk circuits in AT&T's long-distance network. The new system is called *common channel interoffice signaling,* or CCIS for short.

With CCIS, as shown in Figure 5.1, a separate circuit between the offices is dedicated solely as a data link for the transmission of signaling information. No

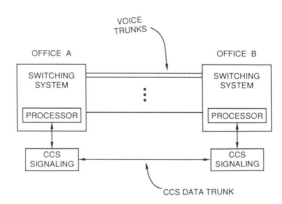

Figure 5.1 Common channel interoffice signaling uses separate, dedicated trunks for signaling. Initially, these CCIS trunks operated at 2.4 or 4.8 kbps over analog trunks. The newer common channel signaling (CCS) uses a dedicated 56-kbps digital trunk for signaling.

signaling information is sent over the voice circuits. The switching machines used in most central offices and in all toll offices are controlled by their own internal digital computers, or processors, that are responsible for the actual switching of the connections. CCIS allows these processors to communicate digitally with each other and to exchange such information as trunk availability and addresses.

A single analog circuit carries the digital data for signaling purposes. Conventional full-duplex modems operating at either 2.4 or 4.8 kbps are used. Each signaling circuit controls about 1800 or 3600 voice circuits, respectively. A newer form of common channel signaling uses digital circuits at 56 kpbs and is described below.

Because the signaling information is transmitted over a circuit separate from the voice circuit, there is a need to determine whether the transmission quality of the voice circuit is acceptable before it is connected for use. This is accomplished by performing a transmission quality check on each voice circuit before it is connected for service. The voice circuit is looped back on itself, and a tone is sent over the circuit. The return level of the tone is checked to be certain it is within specifications.

With conventional signaling over long-distance circuits, a busy tone is generated at the distant office and is sent all the way back to the office near the calling party. This ties up a full voice circuit. With CCIS, a data signal is sent back over the data path, and the busy tone is generated at the near office, hence freeing voice circuits for actual conversations.

The original CCIS system has been enhanced to allow messages (called "datagrams") to be sent to specified destinations in the network switching system. The signaling links are being upgraded to 56-kbps digital circuits between the switching offices. The enhanced common channel signaling (CCS) system is called the common channel signaling 7 (CCS7) network. It is also sometimes known as signaling system 7 (SS7); outside of North America, it is CCITT 17. CCS7 as implemented by AT&T supports integrated-service digital network (ISDN) features, advanced 800 service (such as time-of-day routing), and software defined network (SDN) features. Instructions for network routing along with other information is stored in common databases that are accessed over the CCS network by the switching machines.

Common channel signaling was first installed in AT&T's long-distance network. Common channel signaling is now being installed on a wide scale between local offices where it allows for more efficient use of trunks and also supports such new services as calling-number identification.

CALLING-NUMBER IDENTIFICATION

Although not presently implemented in AT&T's network, common channel signaling has the capability to transmit the calling party's telephone number to the distant office. Some local telephone companies are using their common channel signaling systems to transmit the calling party's number from office to office. This means that the local office knows the telephone number of the calling party. That number can be sent over the local loop to the premises of the called subscriber. This new service is known as *calling-number identification,* or calling-number delivery (CND), or caller ID.

The telephone number of the calling party is sent over the local loop as an in-band signal during the first and second rings of the ringing signal. It is sent as a short burst of digital data at a rate of 1200 to 1800 bps. Phase shift keying is used to encode the digital data for transmission. The digital signal is decoded by circuitry at the customer's premises and is displayed on a unit close to the telephone. The called subscriber either recognizes or does not recognize the number and then decides whether or not to answer the call. Undoubtedly, newer customer-premises equipment will include a directory of the names of frequently received numbers and will display the name, if included in the directory, of the calling party. Perhaps with speech synthesis, the telephone of the future will not ring but will announce the name of who is calling! Of course, that assumes that the person calling from the number is the expected person.

The subscriber to CND usually also receives the capability to block certain incoming calls. Simply by entering a code, calls from specified telephone numbers are forever blocked until the subscriber enters a code to release them. Clearly, crank and annoyance calls are easily traced with calling-number identification. In fact, many telephone subscribers can enter a special code, and the telephone number of the party who just called will be saved and sent to the authorities as an annoying call. Calling-number identification is particularly useful to businesses, since with calling number ID they automatically have the telephone number of the calling party and can use that number to pull up the appropriate customer record from a computerized database.

However, many people are concerned that calling-number ID is an invasion of their privacy. What does the poor spouse do when calling home late at night "from the office" that is really the local bar? These people feel that their telephone number is private and should only be released with their specific permission. Thus, voices of discontent have been expressed toward calling-number ID, and some states have actually issued laws against the service. One solution is to allow the caller to block the forwarding of the calling number to the called party. However, should such blocking be the norm or should it only be done on a specific request basis? Will those who subscribe to calling-number identification simply block all unidentified calls? Will new customer-premises equipment be invented that requires all incoming callers to enter their telephone number or some access code before the called customer's phone will ring?

Calling-number identification, call block, and call trace are only some of a number of new services made possible by the custom local area signaling system (CLASS℠ service). Some other CLASS services are distinctive ringing in which the phone rings in a different pattern for calls from certain preselected numbers, selective call forwarding in which only certain preselected calls are forwarded to another number, and return call in which the last incoming call can be recalled simply by entering an appropriate code. These types of services are sometimes said to be a feature of the "intelligent" network that has evolved because of the use of computers to control signaling.

TELEPHONE NUMBERING PLAN

During the early days of telephony, a total of 10,000 lines was the maximum number of lines served by a telephone exchange. Thus, a four-digit number was necessary to specify the party in a specific exchange. The exchange was specified by two alphabetic characters followed by a single digit, for example WA5 for Waverly-five. Three-digit area codes were later introduced to specify the area of the country to be reached. An area code is more technically called a *numbering plan area* (NPA). Today, we also have country codes for international dialing.

A special nomenclature is used in specifying the overall structure of the telephone numbering plan. The symbol N is used for any of the decimal digits 2 through 9; the symbol X for any of the decimal digits 0 through 9; and 0/1 for the digits 0 and 1 only.

The standard format for telephone numbers in the United States used to be N0/1X-NXX-XXXX. The first three digits, N0/1X, specified the area code or NPA; the next three digits, NXX, specified the local exchange; and the last four digits, XXXX, specified the specific subscriber line in the local exchange. Because the number of area codes possible with the N0/1X format is being exhausted, a new format of NXX is being introduced for the NPA, starting in 1995.

The format N11 is used for special services. Some examples are: 411 for directory assistance, 611 for repair, and 911 for emergencies.

LOCAL-LOOP SIGNALING DESIGN

The flow of direct current along the local loop operates the line relay at the central office, and in this way a request for service is initiated. The amount of current must be sufficient to operate the line relay. However, if the resistance of the local loop is too high, not enough current will flow to activate the relay. The resistance of the local loop depends on the total length of the loop and on the gauge of the wire. The resistance of various gauges of wire is as follows:

26 gauge—83 ohms per 1000 feet,

24 gauge—53 ohms per 1000 feet,

22 gauge—32 ohms per 1000 feet, and

19 gauge—17 ohms per 1000 feet.

The maximum allowed resistance for the local loop can be calculated as follows. The telephone instrument requires about 23 mA of direct current for reliable operation of the carbon granule transmitter. The common battery at the central office has an electromotive force of 48 volts. Thus, the total resistance of the whole circuit (telephone plus local loop plus central office) must not exceed 48/0.023 = 2100 ohms. The resistance of the telephone is equivalent to 400 ohms and the central office equipment likewise has a resistance of 400 ohms. Hence, the resistance of the local loop must not exceed 2100 − 800 = 1300 ohms. This maximum loop resistance of 1300 ohms determines the wire gauge for a given loop length.

REFERENCES

Lawser, John J., and Paul L. Oxley, "Common Channel Signaling Network Evolution," *AT&T Tech. J.*, Vol. 66, No. 3, May–June 1987, pp. 13–20.

Chapter 6
CELLULAR MOBILE TELEPHONE SERVICE

BASIC PRINCIPLES

Cellular mobile telephone service is a high-capacity system for providing direct-dial telephone service to automobiles, and other forms of portable telephone, by using two-way radio transmission. Prior to cellular mobile telephone service, mobile telephony was provided by conventional two-way radio, which allowed only a few dozen two-way radio channels in a given service area. A single, centrally located, high-power radio transmitter served an area about 50 miles in diameter. The service in an area was provided either by a wireline common carrier, such as a Bell company or an independent telephone company, or by a radio common carrier. The very small number of users who could be served in a given area meant that the service was quite costly, and the small capacity of the service meant that many potential customers went unserved. Cellular mobile telephone service solved the problem of congestion, and its mass market made mobile service affordable to many customers.

With cellular mobile telephone service, a 50-MHz bandwidth in the 800- to 900-MHz radio band is used to create 832 two-way radio channels. This alone is a substantial increase over the few dozen channels previously available with conventional mobile telephony. However, the total number of customers served is increased further by reusing channels in the same geographic area. This feat is accomplished by the use of a number of low-power radio transmitters, each serving a small area, or *cell,* within the larger geographic service area. The typical cell has a radius of about 6 to 12 miles. The low power of the transmitter means that the same channel can be used again in another part of the geographic area without causing interference. The whole geographic area is thus divided into cells, with each cell being served by its own transmitter. The cell configuration is chosen to minimize interference caused by the reuse of the channel; hence, the use of the word "cellular."

If congestion starts to appear, cells can be further subdivided, or split, into smaller cells using even lower powered transmitters. Thus, the system can grow gradually to serve more users as demand increases. Larger numbers of users lead to lower prices for the mobile equipment, and this leads to further increases in demand. Cellular mobile telephone service was first made available in the top markets in the United States in 1984, and in a very short time has achieved considerable growth and success. The market now extends beyond automobiles and includes small portable units that can be carried in a pocket. One wonders whether Dick Tracy's wrist radiotelephone is only a matter of a few more years.* The use of two-way radio has now been suggested on a very low power basis to create community systems that could bypass the copper wires of the local loop.

A SHORT HISTORY

Cellular mobile telephone service had lengthy delays in making its way to commercial introduction and availability. These delays were the result of the federal regulatory process. Once it was introduced, however, cellular service became a phenomenal success.

The basic principles of mobile cellular telephone service were formulated at Bell Labs in the late 1940s. The technology to make the service economically feasible, however, was not available until the early and mid-1970s. AT&T had applied earlier to the Federal Communications Commission (FCC) for permission to offer an *advanced mobile phone service* (AMPS) based on the cellular principle, but the FCC wanted to introduce competition into the provision of cellular service. In early 1975, the FCC reallocated a portion of the UHF television band so that it could be used for cellular telephone service. The FCC opened 40 MHz of the 800-MHz radio band to any qualified common carrier, thus bringing competition to cellular service. In March 1977, the FCC granted authorization to the Illinois Bell Telephone Company to install and test a developmental version of AMPS in Chicago. The subsequent test was successful, and in 1983 the Chicago system offered the first commercial cellular service in the United States. The delays in offering cellular service were caused by more than ten years of regulation drafting by the FCC, which finally ended in 1982, in an attempt to determine how to stimulate competition in the cellular area.

The FCC had decided that it would be in the best interests of the public if competition were stimulated in cellular telephony. The FCC implemented this decision by issuing two licenses in each area so that the public would have two service providers from which to choose. The wireline carriers were restricted to one

---

*The character, *Dick Tracy,* created by Chester Gould, is a patented trademark of the Chicago Tribune, Inc. Reg. U.S. Pat. Off. All rights reserved.

license per area. As might be expected, there was an initial flood of applications to the FCC to obtain a license to provide cellular service. The FCC then encouraged competing applications to negotiate some form of consolidation. This took more time and further delayed the introduction of cellular service, particularly for the nonwireline carriers.

Cellular service in each geographic area is provided by a wireline common carrier, which is a subsidiary of the local telephone company, and by a nonwireline common carrier. The radio spectrum space used for cellular service is divided in half to accommodate the two carriers serving a geographic area. The wireline carriers use the so-called B band, and the nonwireline carriers use the A band; each carrier has 416 two-way radio channels, or nearly 60 channels per cell. The mobile equipment itself is available from a number of vendors. This equipment was initially costly, but these costs have dropped substantially as the service has become more widespread.

Cellular mobile telephone service is a great success story. During its first four years in the United States, from 1984 to 1988, it experienced a compound annual growth of more than 100 percent. By mid-1990, its subscribers in the United States numbered 4.4 million. Cellular mobile telephone service is also a great success in Europe, especially Scandinavia, where growth rates rival and in some cases surpass those in the United States. Cellular telephone service was initially targeted at the automobile market, but small portable units have extended the market to nearly everyone on the move with a need to telecommunicate.

SYSTEM OPERATION

A key feature of cellular mobile telephone service is the mobility of the user. This mobility clearly extends across cells, and thus there is a need to track the user and to change radio channels for different cells. This dynamic tracking of the user along with changes in the radio channel is a key feature of cellular service and involves the use of fairly sophisticated technology.

Cells are organized into *clusters.* Clusters are then repeated over the geographic area. As shown in Figure 6.1, the cells can be depicted as hexagons; the actual shape of a cell varies with terrain and radio propagation. In most cellular systems, a cluster consists of seven cells. The 416 two-way radio channels for each system are divided over these seven cells, giving an average of about 60 channels per cell. Cells serving a dense area with many users are usually allocated more channels. The channels are then reused in adjacent clusters. The spatial separation of cells using the same radio channels coupled with their low power reduces any interference to acceptable levels.

Cellular mobile telephone service is a system involving the equipment located at the *mobile unit,* the radio equipment located at the *cell site,* and a *central switch-*

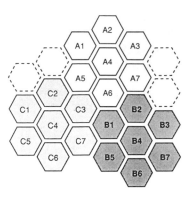

Figure 6.1 Cells are organized into clusters, with most cellular systems using seven cells per cluster. The radio channels are allocated across the seven cells. The clusters are then repeated over and over again to cover the whole geographic area served by the system. Since cells using the same channels are separated from each other, and also since the radio power is low, interference is less likely. Although the cells are depicted as hexagons, their actual shape is quite irregular and depends on the terrain and radio propagation.

Figure 6.2 Photographs of (a) cell site, or base station, and (b) its cellular antenna. (© 1986 Bell Atlantic Mobile Systems. All rights reserved.)

ing office that controls the operation of the whole system and interfaces with the public switched telephone network.

Each cell is served from its own master radio transmitter, receiver, and antenna. The antenna can be either *directional* or *omnidirectional.* An omnidirectional antenna would be situated in the center of the cell, and a directional antenna with a coverage of 120 degrees would be situated at one of the vertices of the cell. A directional antenna is somewhat less susceptible to interference from other cells. The antenna and other equipment located at each cell site constitute a *base station* (see Figure 6.2). Each base station operates in the 800- to 900-MHz range. A radio-

b

Figure 6.2 continued

frequency band from 824 to 849 MHz is used to receive signals from the mobile units, and a band from 869 to 894 MHz is used to transmit signals to the mobile units.

The base station transmits with a power of up to 100 watts, and a nonportable mobile unit transmits at a maximum effective radiated power of about 7 watts. To help control interference, the transmitted power of the mobile unit can be remotely decreased in steps. This is accomplished by a special code that is transmitted to the mobile unit and causes the mobile unit to decrease, or attenuate, its transmitted power to any one of eight prespecified levels. Each level decreases by 4 dB from the preceding level. The code that accomplishes this decrease in transmitted power is called the *mobile attenuation code.*

Because noise immunity is very important, frequency modulation (FM) is used for the radio transmission between the mobile units and the base stations. Narrowband FM with a peak deviation of 12 kHz of the carrier is employed with an rms frequency deviation of 2 kHz for a normal speaker. The radio channels are spaced every 30 kHz in the allocated bands. Since the bands have a total width of 50 MHz, the maximum number of two-way radio channels is 832.

The base stations are connected by land lines to a central place called the *mobile telephone switching office* (MTSO), as shown in Figure 6.3. The MTSO is connected by trunks to the public switched telephone network. An important feature with cellular service is the ability to make telephone calls to the mobile unit over the switched public network as if the mobile unit were a normal telephone with its own 10-digit telephone number. Hence, the MTSO is connected to the public switched network via these trunks.

The mobile unit can move from cell to cell. When this occurs, it needs to change the frequencies of the radio channels used to transmit to and receive from the base stations. This is known as a *handoff.* The relative received signal strength of the mobile unit must be monitored, and when it leaves the range of one base station and enters the range of another base station, information to specify the new frequencies must be sent to it. This information is transmitted as a short burst of data over the voice channel that is being used. The mobile unit needs to know when an incoming call is occurring and also must be able to dial out to set up a call. A *shared data channel,* also called a *paging channel, set-up channel,* or *control channel,* is used for these purposes. The MTSO supervises and controls the entire operation of the system, including the assignment of channels and the changing of frequencies during a handoff. The MTSO sends data over the land lines to the base stations for these control and supervisory functions.

The shared data channel transmits data at a 10-kbps rate. However, data encoding and message repeating reduce the actual information rate to 1.2 kbps, which is equivalent to a practical limit of 25 messages per second when fully loaded. Frequency shift keying is used, with a frequency deviation 8 kHz above and below the carrier frequency. The binary data are encoded using a biphase Manchester

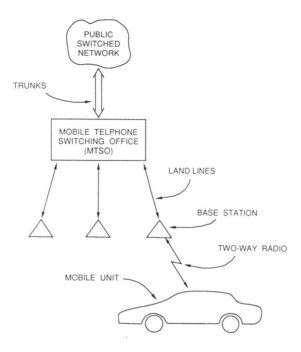

Figure 6.3 The mobile unit communicates by two-way radio with the base station. The base stations all communicate over land lines with the mobile telephone switching office, which then maintains communication over the public switched telephone network.

format, in which a 0/1 transition signifies a binary one and a 1/0 transition signifies a binary zero. The shared data channel is used to request service and to identify the mobile unit when its power is turned on, to assign the initial frequencies to be used for transmission and reception, and to terminate service.

A 34-bit binary number derived from the 10-digit telephone number assigned to the subscriber uniquely identifies each mobile unit and is entered and stored in the mobile unit. This number is sent by the mobile unit to the MTSO where it is used for billing purposes. A 15-bit code entered into the mobile unit identifies the home system from which service is obtained. A 32-bit serial number is factory set into each mobile unit and can be accessed from the MTSO for security purposes.

The 50-MHz-wide radio spectrum available for cellular service is divided into two 25-MHz spaces that are assigned for use by the two separate mobile system operators in each area, a so-called A system and B system, as shown in Figure 6.4. The B system is operated by a subsidiary of the Bell telephone company, and the A system is operated by a nonwireline carrier. The A system uses radio frequencies from 824 to 835 MHz and from 845 to 846.5 MHz for transmissions from the mobile units and frequencies from 869 to 880 MHz and from 890 to 891.5 MHz

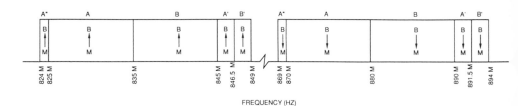

Figure 6.4 Two radio-frequency bands are used for cellular telephone service. The band from 824 to 849 MHz is used for transmission from the mobile unit to the base station and is shared by the A and B systems. The band from 869 to 894 MHz is used for transmission from the base station to the mobile unit and is likewise shared by the A and B systems.

for transmissions to the mobile units. The B system uses 835 to 845 MHz and 846.5 to 849 MHz for mobile transmissions and 880 to 890 MHz and 891.5 to 894 MHz for base transmissions. Transmission from the base station to the mobile unit is called the forward direction, and transmission from the mobile unit to the base station is called the reverse direction. Each of the two systems has a total of 416 two-way radio channels available for its use in these frequency spaces.

Twenty-one channels in the A system and 21 channels in the B system are used as shared data channels or set-up channels. The 21 set-up channels are allocated across the cells, usually with one channel per cell. The mobile unit is programmed to scan all the set-up channels and chooses the set-up channel with the strongest signal. This channel usually corresponds to the nearest base station. The mobile unit can be programmed to scan the set-up channels for only one system or to scan both but with a preference for one system over the other.

A transfer from one radio channel to another is called a handoff. The MTSO makes a decision about whether a handoff is required by measuring every few seconds the strength of the signal received from the mobile unit. This periodic examination of the mobile unit's signal is called *locating,* which refers not to a precise geographic location but rather to signal strength. Thus, a mobile unit could be geographically closer to one base station, but because of terrain, its signal might actually be more strongly received at a more distant base station.

Many business people conduct work from an automobile or while dining at a restaurant by using cellular telephone service. (Photos courtesy of Motorola Cellular.)

Supervision of a call is accomplished by out-of-band tones on the voice channel. A *supervisory audio tone* (SAT) of either 5970, 6000, or 6030 Hz is regularly sent every 0.25 second by the base station to the mobile unit, which then retransmits the tone back to the base station. If the tone is not received back at the base station for a long enough time, it is assumed that the mobile unit has ceased transmitting. The base station sends information (called the *SAT color code*, or SCC) to the mobile unit specifying the specific SAT to be used, and if the actual received SAT disagrees, voice communication is suspended. The round-trip delay of the SAT was at one time suggested as a gross measure of the physical distance of the mobile unit from the base station, but received signal strength is now used to determine handoffs.

A *signaling tone* (ST) of 10 kHz is sent from the mobile unit to the base station. The ST is used to acknowledge any orders from the MTSO, to perform flash requests comparable to the hook-flash of a conventional telephone, and to signal a release request when the user wishes to disconnect the call.

If the channel needs to be changed during a call, information is sent from the base station to the mobile unit over the voice channel as a very brief burst of digital data. The mobile unit decodes this information, changes its transmitting and receiving frequencies to the new channels, and establishes the voice circuit on the new cell. All this happens in a fraction of a second and is usually not noticed by the users.

The data channel is used by the mobile unit to request service. The identity of the mobile unit along with the telephone number of the called party are transmitted over this channel to the base station and then over land lines to the MTSO. The MTSO then establishes a conventional telephone circuit to the called party. When the mobile unit is turned on but is not being used for an actual call, it continuously monitors the strongest data channel for any transmitted paging messages containing its identification number. If the mobile unit recognizes its identification number in any of these paging messages, it responds to the page by sending an appropriate data message back to the base station over the data channel. It then receives information over the data channel telling it what specific voice channel to use to establish voice communication. Once the voice channel has been established, the base station sends a data signal over the voice channel to activate ringing at the mobile unit. The 10-kHz signaling tone is transmitted over the voice channel by the mobile unit to indicate that it is ringing. When this tone ceases, the base station knows that the mobile unit has answered, and the two parties are connected to begin their voice conversation.

A cellular system is designed to handle calls arriving at a mean rate of one call per second in the densest cell. The system can handle one call per subscriber in the busy hour.

Cellular telephony uses high-frequency radio transmission, which is particularly susceptible to reflections that lead to multiple paths from the transmitter to the receiving antenna. Some of these multiple signals arrive out of phase with respect to each other and accordingly cancel. These cancellations cause *fading* of the received signal as the automobile is traveling down the road. Signal processing is used to minimize the subjective annoyance of this fading.

One possible solution to the fading problem is the use of two receiving antennas at the mobile unit. The two antennas are separated by about half a foot. If the signal at one antenna is in a deep fade, then there is a strong probability that the signal at the other antenna is strong. An electronic switch compares the two signals and chooses the stronger. In this way, some of the fading problems caused by multiple-path (multipath) reflections can be corrected. This is know as *switched space diversity* and has been used by at least one manufacturer of mobile units.

INTERSYSTEM ROAMING

A cellular mobile unit will work anywhere there is cellular service, provided that an agreement exists to handle billing back to the unit's home system. A mobile unit could roam all over the country and make telephone calls. Billing is not a problem because the mobile unit contains its own identification number along with an identification of its home system. Reaching the mobile unit is another story, however.

One way to call a roaming mobile unit is for the caller to know the area where the roamer is located and to dial a roamer access number for that area. The caller then enters the 10-digit number for the mobile, and a page is made by that local system. The problem with this method is that the caller must know the roamer access number for the area where the roamer has roamed. A better way is GTE's Follow Me Roaming® service. On entering a foreign cellular area, the roamer sends a special code to the cellular system. The foreign cellular system then notifies the roamer's home system about where the roamer is located. Any calls to the roamer are then automatically forwarded to the foreign cellular system where a page is made to reach the roamer to complete the call.

INCREASING CAPACITY

Cellular telephone service has almost been too successful, and some systems are running out of capacity. Cells can only be split a certain number of times before interference and the handoff rate become serious problems. The question now is one of how to achieve more capacity.

The clearest and simplest solution is to add more spectrum space for cellular radio transmission. However, that space would need to be taken from spectrum space presently allocated to UHF television transmission, but the UHF broadcasters do not want to relinquish any space. Furthermore, some of the systems proposed for high-definition television (HDTV) would use the UHF spectrum. The TV broadcast industry and the cellular industry are engaged in a battle over this issue of spectrum space. My view is that with cable television passing nearly 90 percent of US households along with the strong penetration of VCRs, there is more than enough television diversity and some more UHF space should be made available for two-way cellular service.

Another solution to the capacity problem is the use of technology to increase the capacity while staying within the present spectrum space. Digital has been suggested as one technological method of accomplishing this. In one proposed scheme, three voice signals are digitized at a rate of 9.6 kbps each and are then mutliplexed together in a time-division multiple access (TDMA) scheme. About 10 to 20 ms of each speech signal is stored and then sent in a burst. Three bursts corresponding to the three signals are sent sequentially every 10 to 20 ms within a 30-kHz cellular channel. This digital method triples the capacity of each channel used, and the signal is more secure from outsiders listening to the channel. With present cellular systems, anyone with a radio receiver tuned to the frequency of a cellular channel can listen to conversations.

Systems using a digital bandwidth compression technique called *linear predictive coding* (LPC) have been suggested also. Three voice signals are time-division multiplexed together into a single channel. Each voice signal is LPC encoded at 8 kbps, and an additional 8 kbps is added for handshaking, noise immunity, and synchronization purposes. The three signals require a total of 48 kbps, which can be sent over a single 30-kHz channel.

Another technology being considered to increase the capacity of cellular telephony is *spread-spectrum transmission.* With spread-spectrum transmission, each voice signal is sent over the full bandwidth of the total spectrum space. The trick is learning how to separate the jumble of received signals to concentrate on the desired single signal. One way is to digitize the voice signals and add a unique signature code to each. The receiver then searches for the unique signature code for the desired signal. This is somewhat similar to the way we can listen to one conversation at a cocktail party even though our ears are being bombarded by dozens of speech signals. Another spread-spectrum technique is frequency hopping in which each signal continuously switches channels in a known pattern. Again, the receiver searches for the appropriate pattern for the desired signal. Spread-spectrum transmission can be less prone to interference and thus allows more signals to share the same spectrum space.

A simple way to increase the capacity of a cellular system is the use of single-sideband amplitude-modulation transmission. However, noise and interference problems can be quite severe with AM. The advantage of FM is that it has a capture effect in that the strongest FM signal is locked onto by the receiver, and all weaker interfering signals are ignored. It is quite conceivable, however, that ways will be invented to improve on conventional FM and AM techniques to increase the capacity of a cellular system without creating serious interference problems.

Whatever happens to increase capacity, the problem of standardization looms ahead. The present analog FM scheme is now established as the standard for millions of cellular users. The challenge for the cellular industry is to choose a new standard and then migrate toward it while still supplying service to all the users of the older technology. This is no easy task!

CELLULAR TELEPHONY—AN ASSESSMENT

Cellular telephony involves very sophisticated technology. The mobile unit is capable of being tuned automatically to any one of 832 two-way radio channels and can transmit and receive binary data and act on that data. The mobile telephone switching office must monitor thousands of calls, determine when handoffs are needed, send and receive data transmissions from the mobile units, and determine billing. It is astonishing that mobile units have so shrunk in size that portable cellular phones that fit in a pocket are readily available.

Yet more important than all the technological sophistication is the ease of use of cellular telephony. When one is driving along a busy freeway, bumper to bumper, at 55 miles per hour, it must be easy to place and receive a cellular call. The dialing buttons must be well placed for ease of use, without taking one's eyes off the road ahead. The challenge for future cellular phones for the car is the user interface. This is one area where synthetic speech could augment the visual display of a telephone number. In cellular dialing, the user enters the number into the mobile unit, and only after the complete number has been entered does the unit send the number to the base station. The user can review the dialed number on a visual display before sending it out, but a synthesized-speech "readout" of the number would perhaps be most useful during driving. So too would be speech recognition so that the user could speak the digits of the number to be called.

Cellular mobile telephone service continues to be very exciting and have a great growth potential. Some futurists predict a day when we will all carry our own personal telephone, served by some form of cellular service. A form of local or neighborhood two-way radio service, called a personal communication network (PCN), already has been proposed. There are also cordless telephones that work over about 100 feet and enable users to walk about the home talking on the telephone. These cordless phones use two-way radio and connect to the standard telephone line. With portable phones, cellular phones, and cordless phones, we have the ability to be reached and to reach anyone by telephone anywhere at any time. Whether we will all want to be that easily reached at all times, though, remains to be seen.

REFERENCES

"Advanced Mobile Phone Service," *Bell Syst. Tech. J.,* Vol. 58, No. 1, January 1979.

"EIA Interim Standard No. IS-3-D for Cellular System," Washington, DC: Electronic Industries Association, March 1987.

Prentiss, Stan, *Introducing Cellular Communications,* Blue Ridge Summit, PA: Tab Books Inc., 1984.

Chapter 7
DATA COMMUNICATION

DIAL-UP TELEPHONE NETWORK

Vast amounts of information are stored in computerized data banks, or *databases,* such as credit histories, airline schedules, and law cases. This information can be accessed remotely using either a data terminal or a personal computer by simply establishing a telephone connection to the database. Thus, the switched telephone network is the most ubiquitous way of communicating data. However, computers use binary signals—in the form of binary digits, called *bits*—for communication, but the telephone network has been designed for analog, speech-like signals. Hence, the binary signals required by the computer are converted to analog tones that can be conveyed over the telephone network.

The tones are *carrier signals,* and certain of their parameters are varied according to the binary data. The process of causing these variations is called *modulation,* and the process of extracting the binary data is called *demodulation.* A device is needed between the computer and the telephone line to perform the translation between the binary data required by the computer and the modulated tones required for transmission over the telephone network. Because communication is a two-way affair, this device must perform both modulation and demodulation of the tone, depending on the direction of the communication. Such a device is both a modulator and a demodulator, and thus is called a *modem* for short. A data terminal usually has a modem built into it.

By way of a modem, any computer can use the dial-up telephone network to reach most computerized database services. The speed of the communication can be as low as 300 bps to as high as 19,200 bps, with 1200 and 2400 bps being popular intermediate speeds. The 4-kHz bandwidth of a telephone circuit along with the signal-to-noise ratio limit the data capacity of the dial-up telephone network. Most modems allow simultaneous, two-way communication—a type of communication that is called *full duplex.* A type of communication that is one-way at a time and requires a reversal of the direction of the communication to achieve two-way com-

---

munication is called *half-duplex*. A strictly one-way type of communication, such as television broadcasting, is called *simplex*.

The dial-up telephone network is ubiquitous, easy to use, and very economical for much data traffic. However, it does have some problems for data communication. Longer local loops have loading coils, and these loading coils restrict the higher frequencies that can be carried over the loop, thereby restricting the upper limit on the data capacity. For certain types of high-speed data communication, the loading must be removed. The echo suppressors used on long-distance circuits can play havoc with full-duplex data communication and must be removed for certain types of full-duplex data communication. The echo suppressors do not allow simultaneous two-way communication, and the time required for them to switch the directionality of the circuit is too lengthy for the high speeds of data signals.

ASCII

Most of the information sent as data over communication networks is text and alphabetic-numeric characters. These characters are encoded as binary information using the American Standard Code for Information Interchange (ASCII). ASCII uses seven bits to represent a total of 128 possible characters, including uppercase and lowercase letters, numbers, special control symbols, and such characters as /?&*%#. An extra eighth bit is added as an error-detection bit. It is set to either a zero or a one to make the total number of ones in the eight-bit string either an odd or an even number (called odd parity or even parity). ASCII is referred to as an eight-bit code. Figure 7.1 shows the ASCII transmission for the character N.

Some data terminals transmit a single ASCII character at a time. The receiving equipment does not know when a transmission might occur, and hence the idle state of the transmission must be changed to alert the receiving equipment that a character is about to be sent. The idle state usually is a continuous transmission of

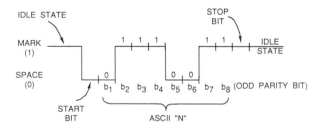

Figure 7.1 The character N is encoded in ASCII as a series of eight bits represented as marks and spaces. The signal arrives left to right, with the first arrival being a start bit to alert the receiving equipment to expect a character. Next come the 7 bits for the ASCII character, with the eighth bit being an odd parity bit. A stop bit indicates the end of the transmission.

ones, called *marks,* so that any break in the line can be ascertained. Thus, the first bit sent to alert the receiving equipment is a zero, called a *space.* This is followed by the eight bits for the ASCII character. The least significant bit b_1 arrives first followed by the other seven bits with the parity bit arriving last. Some time must be allowed for the circuit to return to the idle state, and hence a mark is sent at the end as a so-called stop bit.

This form of data transmission in which a single character is sent at a time is called *asynchronous transmission.* Asynchronous transmission is wasteful of capacity because a start bit and a stop bit must be sent for each character. A more efficient form of data transmission is synchronous transmission, in which a whole block of characters is sent in one long transmission. Some special codes start the transmission to synchronize the receiving equipment to the precise rate of transmission. This type of transmission is found in message and packet switching.

MESSAGE AND PACKET SWITCHING

The telephone network is a line-switched, or *circuit-switched,* system. A complete, dedicated connection is first established, and then actual voice conversation begins. The connection is maintained for the full duration of the call, even if no one is speaking. This type of system is fine for speech communication, but it entails numerous problems for data communication.

Unlike speech communication, interactive data communication consists of short bursts of information with a fair amount of dead time between each burst. The circuit-switched telephone network in which a complete connection is maintained for the duration of the call is very wasteful for this type of bursty data communication. Humans are willing to dial a telephone number and wait seconds for the connection to be established. Computers demand instant access and cannot wait idly for a circuit connection. Short busts, instant access, and long holding times characterize interactive data communication.

Message switching and *packet switching* are appropriate for data communication. With message and packet switching, the intended destination is added to the data to be sent. The data along with its destination are gradually switched, stored, and forwarded through the data network to the final destination. This type of switching is also called *store-and-forward* communication.

In general, a *message* is one long burst of information. This long burst of information may be broken up into shorter bursts of fixed lengths, called *packets.* As shown in Figure 7.2, packets may be switched and transmitted over a number of different paths before reaching the final destination, where they are reassembled into the complete message. With message switching, shown in Figure 7.3, the message is switched and transmitted in its entirety over the data network. The message may be stored temporarily for some time along its transmission while waiting for an available circuit with sufficient capacity. A data packet includes both the *desti-*

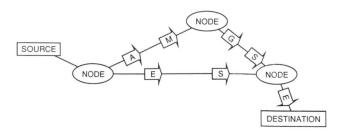

Figure 7.2 With packet switching, a complete message is broken into a series of shorter, fixed-length packets. The packets travel separately along the network from their source to the destination. The packets are assembled at the destination to form the original, complete message.

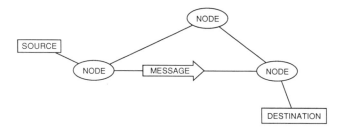

Figure 7.3 With message switching, a complete message travels as a whole along a data network from its source to the destination. The message may wait at nodes along the way for sufficient transmission capacity.

nation address and the *source address.* The actual data are typically about 1000 bits.

A packet-switched data network has nodes at which computers examine the packets and then send them along toward their final destination. Different paths may be taken by different packets depending on the available transmission capacity between the nodes. Digital circuits form the transmission capacity between the nodes. Depending on the traffic, the digital capacity could be as small as a single 64-kpbs circuit, a 1.54-Mbps DS1 circuit, or even higher.

PROTOCOLS AND STANDARDS

When we speak over the telephone, certain agreed-on rules and procedures govern our conduct. We answer the telephone by saying "hello." We wait for a pause in the other person's conversation before interrupting. At the end of our conversation, we say "good-bye." Similarly, we require agreed-on rules and procedures to ensure the orderly transfer of data between digital devices that are interconnected by communication facilities. These rules and procedures are called data communication *protocols*. The manner in which a terminal acknowledges that it is ready to receive data is an example of a protocol.

Many *interfaces* are encountered in data communication, such as the connection of the terminal to the network and how information displayed on the screen applies to a specific application. Clearly, interfaces and protocols must be standardized or total chaos would result. To add some sense of order to the many interfaces, protocols, and standards encountered in data communication, the International Organization for Standardization (ISO) has created a conceptual model for data communication consisting of seven separate levels, or layers. Since the model is intended to facilitate the open interconnection of data systems, it is called the open systems interconnection (OSI) model. The electrical connection between a modem and a terminal is an example of one of the interfaces considered by the OSI model for which standards are specified.

Although the seven-layered OSI model is intended to eliminate confusion, for me the seven layers and what they do is complex and confusing. The lowest layer (layer 1) is concerned with hardware and its physical and electrical connection. The highest layer (layer 7) is concerned with the specific application of the communication, for example, order entry or funds transfer. The seven layers are:

1. Physical layer,
2. Link layer,
3. Network layer,
4. Transport layer,
5. Session layer,
6. Presentation layer, and
7. Application layer.

The first three layers (physical, link, and network) are concerned with routing of data from one piece of equipment to another. The last four layers are concerned with the actual dialogue for a particular application. The user interacts with the highest layer, the application layer. Information is then transferred from layer to layer and finally across the communication network to the receiving end where the information ascends to the application layer at the host computer. Each layer packages the information differently, but the contents are unchanged. The "user" may be a computer program in some applications.

As mentioned, the three lower level layers apply to the method of data transport through the communication network. The physical layer is concerned with the physical and electrical interconnection of hardware. For example, the type of coding used to represent a bit, for example, by a change in voltage, would be specified in the standards for the physical layer. The physical layer is concerned with the cable and wires used to interconnect data equipment, such as the RS-232-C standard for specifying the various voltage levels and signals on a 25-pin connector. The link layer is concerned with groups of bits, called *frames*. The use of checksums for error detection and correction is specified at the link layer, for example, along with protocols for acknowledging the receipt of data. The use and placement of the address for a data message sent over a packet-switched network is specified in the protocols for the network layer. The network layer is concerned with the actual routing and transmission of the data through the network. For a packet-switched network, the rules governing the actual routing of packets from node to node and to the final destination are specified in the standards for the network layer.

The four higher level layers are concerned with the specific application and its software. The applications layer, for example, specifies the task or job that the user is accomplishing, for example, videotex, electronic mail, or funds transfer. The presentation layer is concerned with such areas as the encryption of data, the positioning of the cursor on the display screen, the encoding of graphical symbols for display, and the standards needed for representing various types of data files. The time between when a user first enters a computer application ("logs in") and when the user finishes the application ("logs out") is called a *session*. The session layer is concerned with session-specific areas, such as user identification and recovery from transport failure. The transport layer is concerned with such areas as how data are handled when errors are detected and how to handle data overflows from network congestion.

Standards are available for most of the layers, but engineers continue to refine the definition of some of the intermediate layers and their appropriate standards and protocols. IBM has its own layered data networking protocol, called System Network Architecture (SNA), and Digital Equipment Corporation (DEC) has its own, called Digital Network Architecture (DNA). The implementation of DNA is called DECNET. One of the earliest packet-switched networks was ARPANET, developed by the US Defense Department's Advanced Research Projects Agency. ARPANET was originally developed to allow large computer systems to share resources; the actual use was to send messages, or electronic mail, between the computer users.

Information for data transmission over specialized data network links is organized into a grouping called a *frame*. The beginning of a frame has a *header* containing address information and a *trailer* to indicate the end of the frame. A large block of characters forms the message to be transmitted, and the header and trailer are appended at the link layer. One popular protocol for this type of transmission is IBM's Binary Synchronous Communication, also known as BSC or as BISYNC. A newer IBM protocol is the Synchronous Data Link Control (SDLC), which accommodates any number of data bits and includes sophisticated error checking.

The use of many different protocols for different types of databases and computers can be a problem. One solution is to fix a place in the network where conversion from one protocol to another can occur. These protocol conversion places are called *gateways,* and they interconnect different data networks, databases, and computers.

LOCAL AREA DATA NETWORKS

The widespread use of personal computers as workstations in offices, schools, and other places has created a need to share certain limited, expensive resources, such as laser printers and large disk memories. To accomplish this sharing, a number of computers must all be able to access the same equipment. This can be done through the use of cables and a switch of some kind, but a more flexible way is through the use of a local data network linking together the computers and the shared equipment. Such networks are called *local area networks,* or LANs. A wide variety of LANs are available, depending on such factors as the network *topology,* the transmission *medium,* the *multiplexing* technique, and the *access control* protocol. Most LANs use some form of packet switching.

The various network topologies are the *star,* the *ring,* the *bus,* and the *mesh;* these topologies are shown in Figure 7.4. In the mesh, each node is connected directly by separate wires to all, or several of, the other nodes. A switch at each node makes the needed connection for the transmission of data. The mesh topology needs a lot of wire for a large number of nodes, and thus is impractical in most cases. The star topology has a centralized hub and switch to which all nodes are connected. The star topology is similar to the local loops and central office used for telephone communication. The mesh and star topologies are primarily of academic interest because they are not used for most LANs.

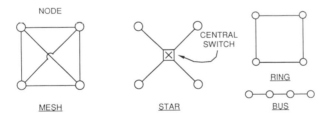

Figure 7.4 Various topologies for networks are the mesh, the star, the bus, and the ring. The last two are used frequently for local area networks.

The ring and the bus topologies are similar in that control is decentralized and the same data signal is broadcast to all nodes. The ring is a loop of some appropriate transmission medium that connects from one node to the next. Data signals are passed or circulate around the loop and are usually amplified and repeated at each node. The bus is a single length of some appropriate transmission medium.

Each node is linked to the medium at connections called *taps*. A bus can be visualized as an opened or broken loop.

Many different transmission media are used for LANs. The transmission medium can be twisted-wire pair, coaxial cable, or optical fiber. Both coaxial cable and optical fiber require special connectors at the places where the nodes are attached. Twisted-wire pair, on the other hand, is very simple to connect at the nodes and also is the least expensive and the easiest to install.

The data capacities of optical fiber and of coaxial cable are much greater than twisted-pair wire. Coaxial cable and optical fiber can carry hundreds of megabits per second, while twisted-pair wire can carry only a few megabits per second. Clearly, the choice of transmission medium must consider the data capacity that is needed. A full page of text, for example, is equivalent to about 40,000 bits. If this page is sent to a high-speed laser printer operating at one page per second, then a transmisison capacity of less than 50 kbps will suffice, and 100 Mbps would be sure overkill. However, if a number of users are sharing complex graphical information, then transmisison capacities of the order of megabits per second indeed make good sense. The capacities of the LAN must be matched to the needs of the users.

With the ring and the bus topologies, all nodes share the same medium, and the signals must somehow be kept separable through appropriate multiplexing technique. The medium can be shared in time through time-division multiple access or in frequency through frequency-division multiple access. Time-division multiple access is most frequently used for LANs.

With time-division multiple access, bursts of data can be transmitted over the full bandwidth of the medium in very short intervals of time. A node that has data to be sent, however, needs to know whether the medium is available for use. Various techniques have been devised for controlling this access to the medium. Each node may be polled in turn to determine whether it has data to be transmitted. The problem with this polling method of access control is that some form of central control is needed to perform the polling. Nodes may reserve time slots for the transmission of their data, but here too some form of central control of the reservations is needed. The problem with central control is that it adds to the expense and complexity of the LAN, and hence some form of self-control is preferable.

A way for self-controlled access is for each node to check the status of the medium and, if it is not in use, to send the data. Data are sent by modulating a carrier, and hence the status of the medium can be determined simply by determining whether a carrier is or is not present. This type of multiple access by sensing a carrier is called *carrier-sense multiple access* (CSMA). The problem is that *contention* can occur if two nodes try simultaneously to send some data, but procedures have been devised to resolve this problem.

In one procedure for resolving contention, if the medium is already in use, the node with data to be sent simply waits. The node determines whether the medium is already in use by sensing the presense of the data carrier on the medium. However, two nodes might separately sense no carrier and then simultaneously begin transmission. If this happens, a data *collision* occurs, and both nodes sense that some other node is transmitting too. The protocol specifies that both nodes cease transmission and transmit later. An algorithm is used to determine when to attempt retransmission; a random element in the algorithm reduces the probability that both nodes will wait identical times. This access protocol is called *carrier-sense multiple access with collision detection* (CSMA/CD). The CSMA/CD protocol is used with Xerox's Ethernet® LAN, which has a bus topology using coaxial cable as the transmission medium.

A second protocol involves a *token* that is circulated around the network. The "token" is a short data message. If a node has data to transmit, it must possess the token in order to transmit. The node must wait for the token to come into its possession. After transmitting its data, the node releases the token. Alternatively, the node may change the contents of the token to signify that data transmission is occurring and then recirculate the reserved token. Changing the token back to its free state releases the token for use by another node. *Token passing,* as this protocol is called, is particularly appropriate for the ring topology (called token-ring), although it is used with the bus topology too (called token-bus).

With frequency-division multiple access, channels are dynamically assigned and switched to connect those nodes that are transferring data. This assignment of channels is performed by a master controller. Some nodes may share the same channel, and in this case, some form of contention-resolving procedure is needed. The use of frequency-division multiplexing to share a medium is called *broadband transmission.*

If a medium is shared in time through time-division multiple access, the data signals are usually transmitted as *baseband* digital data. The terms "baseband" and "broadband" are unfortunate because a medium such as coaxial cable has a broad bandwidth that can be used with either baseband or broadband transmission.

Computers, terminals, printers, and other devices are usually connected to a LAN through two *interfaces,* as shown in Figure 7.5. One interface is the connection to the network itself through a *network interface* unit. For a bus network, the network interface could be as simple as an electrical connection to the line by a parallel connection, called a *tap.* For a loop network, the network interface receives signals sent along the loop, extracts the portion intended for the node it serves, and retransmits the signal along the loop. The network interface unit for a loop network performs as a transmitter-receiver, or *transceiver.*

The second connection is to the device itself. If the device is a shared resource, such as a printer or a hard-disk file, the interface must manage and serve the access by a number of users. It is thus called either a *printer server* or a *file server,* respectively. If the device is a workstation of some kind, such as a personal computer, the interface is called a *station interface,* or a *terminal server,* and is usually a printed circuit board, or card, that plugs into the station itself. Servers and station interfaces are microprocessor-based computers that perform communication management functions.

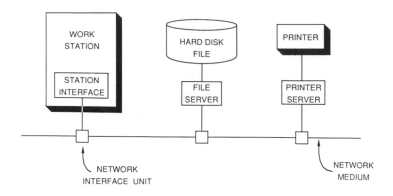

Figure 7.5 Computer workstations, disk files, and printers are connected through two interfaces to a local area network. One interface is the physical connection to the transmission medium of the network. The other interface manages the data traffic and is located at the device being served.

ISDN

We saw in earlier chapters that voice is routinely converted to a digital signal for transmission and switching over the telephone network. In this chapter, we see that data are already in the form of a digital signal. The synthesis of voice and data on an all-digital communication network thus seems to be the normal progression. This same network could also carry digitized video and digital facsimile signals because, in digital form, all signals are simply bits. Different services require only different bit rates. The overall concept of such an integration of services through an all-digital network is called *integrated services digital network,* or ISDN for short.

ISDN ultimately is a worldwide network composed of many smaller private and public networks, all interconnected through common standards, and providing a total end-to-end digital capability. This end-to-end capability would extend directly to a standard digital outlet on each customer's premises. At a minimum, three two-way digital circuits would be offered in the following configuration: two B channels each operating at 64 kbps and a single D channel operating at 16 kbps. This minimum configuration is known as the 2B+D basic rate, or the *basic rate interface* (BRI). The B channels would be used for digital voice and for high-speed data communication, perhaps even simultaneous voice and data communication, and the D channel would be used for signaling and lower speed data communication. Telephone instruments with built-in analog-to-digital converters would be needed to connect to the basic access interface. The existing copper wire in the local loop could, in most cases, accommodate this basic access to ISDN.

A second configuration would consist of 23 B channels, each operating at 64 kbps, and a single D channel operating at 64 kbps. This configuration is known as the 23B+D primary rate, or the *primary rate interface* (PRI). Twenty-four 64-kbps digital circuits multiplexed together is the standard DS1 or classic T1 digital signal. Thus, the backbone of ISDN is the standard DS1 signal at 1.544 Mbps. The European version of ISDN would offer 30 B channels and 2 D channels as their primary rate at 2.048 Mbps.

Another form of ISDN that is talked about today is *broadband ISDN* in which a digital signal at rates as high as several hundred megabits per second would be brought to the customer's premises. This broadband capability would be used for such services as high-definition television and two-way picturephone service. More is said about these futuristic services in a later chapter.

The primary driving force for ISDN is technological in nature. Indeed, the network is primarily digital today, but the local loop is still primarily analog. Why then should the digital capability in the network be extended along the local loop to the customer's premises? To answer this question, we need to examine the services that would be made possible with ISDN. The data services most frequently mentioned are electronic mail for the business market and videotex for the home market. Most large businesses already have their own PBX, which most likely is a digital machine. The home market for videotex is very doubtful, and billions of dollars have already been lost attempting to promote that market. Data communication in many businesses is for the most part local and is carried over LANs. Such services as facsimile work over existing analog circuits. Picturephone has failed in the past, and there is no reason to expect that its success would somehow be guaranteed by ISDN. We thus see that from a consumer perspective there is little need for ISDN. Thus, ISDN appears to be a technological solution in search of a problem!

Indeed, the network is digital, and in digital form, all signals are the same. Thus, if one defines ISDN as a digital network carrying digital signals representing many different services, then ISDN is already here. But if by ISDN one means an end-to-end digital capability over the local loop, then ISDN is a long way from arriving. In the next century, optical fiber will replace copper wire in the local loop on a grand scale. This will occur for engineering and economic reasons for the delivery of telephone signals and not for the delivery of new services. Clearly, the signals carried over the fiber will be digital, and then ISDN will have arrived in its end-to-end definition.

VOICE *VERSUS* DATA

The information age is upon us, and some futurists predict that the data traffic generated from such services as electronic mail, database access, and electronic funds transfer will soon outstrip voice traffic. Let us examine this astonishing proposition to see whether it is realistic or preposterous. The examination will estimate the yearly total traffic in bit equivalents for various communication services.

The average daily number of telephone conversations in the United States in 1988 was 1700 million (*Stat. Abs.,* 1990). Assuming that the length of an average telephone conversation is three minutes and that the speech is encoded into digital form at 64 kbps in each direction, we have a total yearly traffic of 14×10^{18} bits, or 14,000 quadrillion bits (a quadrillion is 10^{15}). We next compute the traffic from various text-based communication services.

In 1988, 93.2 billion pieces of first-class, second-class, and priority mail were carried in the United States. Assuming that each piece of mail contained the equivalent of one full page of text (4800 characters at 8 bits per character), we have a total yearly traffic if all this mail were communicated electronically in digital form of 3.6×10^{15} bits, or 3.6 quadrillion bits. There were 860 million credit cards of various kinds outstanding in 1988. If each card were used twice a day and each transaction required 200 characters, we have a total yearly traffic of 1.0×10^{15} bits, or 1.0 quadrillion bits. If we assume that each of the 93 million households in the United States generates or receives two checks a day and that each check if communicated electronically in digital form requires 200 characters, we have a total yearly traffic of 1.1×10^{14} bits, or 0.1 quadrillion bits.

Clearly, these services do not come even close to the traffic generated by voice telephony. We will assume that the information age has truly arrived and that each of the 122 million workers in 1988's labor force works at terminals, entering and retrieving information all day, 365 eight-hour workdays a year. We assume further that they are very productive and generate or use 30 full screens of information per hour, where a screen consists of 1500 characters. This information-crazed workforce would generate a yearly total of 1.3×10^{17} bits, or 130 quadrillion bits. Telephony still wins!

One service that could come close to voice telephony in traffic is the electronic delivery of newspapers. The average length of a newspaper is 100 pages. We will assume that each page contains about 7000 characters of text (56,000 bits) and about 200 square inches of pictures. Using a resolution of 200 lines per inch and 20:1 compression, this pictorial information is equivalent to 400,000 bits per page. The transmission in electronic form of such a newspaper per day to the 63 million newspaper readers in the United States would generate a total yearly traffic of 1.0×10^{18} bits, or 1000 quadrillion bits.

The results of the preceding estimations are:

SERVICE	NUMBER OF BITS
Telephony	$14,000 \times 10^{15}$
Mail	3.6×10^{15}
Credit cards	1.0×10^{15}
Checks	0.1×10^{15}
Information workforce	130×10^{15}
Electronic newspaper	1000×10^{15}

Clearly, voice telephony swamps all other communication services. This really is not that surprising since each telephone conversation is a unique message requiring a considerable amount of bits. If speech were converted to text, considerable savings in bits would result, as shown in the following section.

SPEECH AND TEXT

A fairly good rate for reading aloud is about 120 words per minute. If we assume that an average word is six characters long, then the text equivalent of one minute of speech is 720 characters, or 5760 bits using eight-bit ASCII coding. This one minute of speech requires 3,840,000 bits if encoded digitally at 64 kbps. Clearly, text is more efficient than speech in its use of bits by a factor of nearly 700 to 1.

Voice mail is a form of electronic mail in which the actual speech message is stored in digital form for later retrieval by the intended recipient. The digitalization of speech at a rate of 64,000 bps requires large amounts of storage. As shown above, each minute of speech requires 3.8 million bits. However, there is considerable redundancy in the speech wave, and appropriate methods have been developed to reduce the bit rate for storing speech to 9.6 kbps, and less. If we assume that the speech can be digitized and stored at a rate of 1.2 kbps, then a one-minute message would require only 72,000 bits. This, however, is still more than 10 times the storage needed for a text message in ASCII. Thus, text is more efficient for storing messages, even if compression of the speech signal is used. Yet we all have telephones—we do not all have text terminals—and nuances of inflection are lost in text messages. Regardless of storage efficiency, voice mail is here to stay.

TEXT AND IMAGE

Electronic mail (E-mail) entered from a keyboard and sent as a data signal to the recipient is used frequently in many large businesses. However, the considerable growth of facsimile during the last few years indicates that an easy-to-use, standardized system based on the public switched network is impossible to beat, even though E-mail is more efficient in its use of bits, as we shall soon see.

We assume a full page of text with 60 rows of 80 characters each. These 4800 characters coded in ASCII require 38,400 bits. This same page of text is scanned by a facsimile unit with a resolution of 200 lines per inch, truly excellent quality. The 8.5- $\times$ 11-inch page has about 3.7 million picture elements (pels), and if a single bit is used to encode each pel, then about 3.7 million bits are needed for a facsimile of the page of text. However, there is a great deal of empty space on a page of text, and many patterns appear again and again. Hence, a large amount of compression is feasible for facsimile reproduction. Compression by 15 to 1 is quite possible, thereby reducing the number of bits per page to about 250,000 bits—6.5 times that needed for the ASCII representation. We must remember, though, that facsimile reproduces a handwritten signature, letterhead graphics, and pictorial information—ASCII does not.

A PICTURE IS WORTH 1000 WORDS

In analog forms, a speech signal has a bandwidth of 4 kHz, and a video signal has a bandwidth of 4.5 MHz—a ratio of about 1000 to 1, thereby proving the old adage that a picture is worth 1000 words!

There is considerable redundancy in a video signal, both within an individual frame and also between frames. This redundancy can be used to reduce the bit rate required for a video signal in digital form. The digitalization of a video signal can result in a data rate of 90 Mbps, as opposed to a digital speech signal at 64 kbps. The digitized video signal can be compressed to 1.5 Mbps with some degradation of quality, particularly for those portions of the video with considerable motion. The speech signal can also be compressed, perhaps to a data rate as low as 1.2 kbps if some degradation in quality is acceptable. In compressed forms, our ratio of 1000 to 1 is still preserved.

VOCODERS

There is considerable redundancy in a speech signal, and methods have been devised to reduce the analog bandwidth and digital data rate needed to transmit a speech signal. These reductions are performed by voice coders and decoders, called *vocoders.* Human speech is analyzed and converted to parameters that require less analog bandwidth or fewer digital data rates for transmission. At the receiver, the received parameters are used to construct a synthetic speech signal.

The parameters used to encode the speech signal are derived according to which model is used to represent the speech. One model represents the speech signal according to its spectrum. A vocoder based on this spectral representation is called a *channel vocoder.* A representation of the source signal is also needed along with a determination of the pitch of the voiced intervals of the speech signal. A channel vocoder can transmit a telephone-quality speech signal in 300 Hz, or in digital form, about 2000 bps. This is a considerable savings over the 3500 Hz required in analog form or the 64,000 bps in digital form.

Other models are used to represent the speech signal. The formant model represents the spectral information in terms of the resonant peaks, or *formants,* in the spectrum. The articulatory model represents the speech in terms of the positions of the tongue and lips. The popular linear predictive coding model represents the speech signal in terms of the parameters of a digital filter. With the LPC technique, the speech signal can be compressed to 1200 bps.

One problem with all of these techniques is the cost and complexity of the coding and decoding equipment required for each voice circuit. The ever-increasing bandwidth of transmission systems simply has outpaced the vocoder technology. There is one exception, however, and that is the use of speech coders in voice encryption for security purposes. Perhaps the biggest problem with speech coding is that the synthesized speech is not natural sounding and frequently has a buzzy quality. However, a fair amount of research is being conducted to solve this problem.

REFERENCES

Davenport, William P., *Modern Data Communication,* Rochelle Park, NJ: Hayden, 1971.

DeNoia, Lynn A., *Data Communication,* Columbus, OH: Merrill Publishing Company, 1987.

Doll, Dixon R., *Data Communications,* New York: John Wiley and Sons, 1978.

FitzGerald, Jerry, *Business Data Communications,* New York: John Wiley and Sons, 1984.

Stallings, William, "A Manager's Guide to Protocols for Local Networking," *Telecommunications,* September 1987, pp. 38–54.

Chapter 8
A SHORT HISTORY OF THE BELL SYSTEM

Almost as important as the invention of the telephone was the invention of the Bell System to provide telephone service. The structure of the Bell System survived for nearly a century before its fragmentation in the mid-1980s. The forces that led to this fragmentation were technology, competition, and government antitrust action. What is very enlightening is that these three forces were present, in various forms and degrees, from the earliest days of telephony and shaped the Bell System throughout its history. Thus, an understanding of the history of the Bell System can help us understand more fully the present and perhaps even the future of telecommunication.

It is impossible for one chapter to do justice to all the details and nuances of the 100-year history of the Bell System. The history then in this chapter highlights key events that will help us understand the recent and continuing telecommunication policy questions of today.

Alexander Graham Bell is credited with the invention of the telephone. Bell, born in 1847 in Edinburgh, Scotland, came to Canada with his parents in 1870. In 1872, he moved to Boston where he taught teachers of the deaf along with his own deaf pupils. Bell dabbled with early telegraphy and became fascinated with the idea of sending speech directly over telegraph wires. Bell and his colleague, Thomas Watson, worked in Boston to perfect his idea of sending speech electrically, and on February 14, 1876, Bell applied for a patent on this idea. The patent was allowed on March 3, 1876, and issued four days later on March 7. The first working demonstration by Bell and Watson of the telephone did not occur until a few days after that, on March 10, 1876.

Controversy clouds the invention of the telephone. On February 14, 1876, the same day that Bell filed his patent application, Elisha Gray, an electrical researcher and a founder of the Western Electric Manufacturing Company, filed a caveat, or warning, to other inventors concerning a speaking telephone on which he was working. Gray's disclosure included a key concept of a variable-resistance transmitter and was filed only a few hours after Bell's patent application. Bell's application also disclosed the idea of a variable-resistance transmitter, but this idea was written by hand in the margin of the application, almost as if an afterthought. Whether Bell somehow heard of Gray's idea and then added the marginal notation to the application will never be known. In a split decision, the US Supreme Court upheld Bell's claim, but Gray died convinced his ideas were stolen from him by Bell. In addition to his pioneering work in telephony, Gray perfected the telautograph for transmitting handwriting to a distant point, and the Gray National Telautograph Company was chartered in 1888 to pursue his invention.

Bell's teaching of the deaf was closely tied to his financial backers, Gardiner Greene Hubbard and Thomas Sanders. Hubbard had a deaf daughter, who would marry Bell on July 11, 1877. Sanders had a deaf son who was Bell's pupil. On July 9, 1877, the Bell Telephone Company of Massachusetts was formed by Hubbard, Sanders, Bell, and Watson to promote the telephone commercially. Bell's telephone was an instant success and attracted a considerable amount of favorable attention. In the fall of 1877, the early Bell Company had 600 subscribers to their private-line service. By the end of the year, they had more than 3000 subscribers.

Competition almost immediately characterized the early days of telephony. In December 1877, the Western Union Company created the American Speaking Telephone Company to sell telephone service, utilizing technology invented by Elisha Gray and Thomas Alva Edison. Western Union claimed that their telephone service offered better quality than Bell's.

Early telephone service was offered only as private-line service. That changed when in early 1878 exchange telephone service was offered for the first time, and a telephone could reach others within the same exchange. However, the Bell telephones were not interconnected with the Western Union telephones, and hence two exclusive telephone systems existed in each area—not a particularly efficient way of providing telephone service.

Bell needed more funding for additional expansion, and in June 1878, the Bell Telephone Company of Massachusetts was reorganized to create the Bell Telephone Company. The Bell Telephone Company granted licenses to other companies to install and sell telephone service in all places other than New England, which was the province of Bell's New England Telephone Company formed earlier that year. As of July 1, 1878, there were 10,755 Bell telephones in service. The early years saw many such reorganizations for the purpose of infusing new capital into the rapidly expanding telephone business. A person later to play an essential role in the formation of the Bell System, Theodore N. Vail, was hired at 33 years old as the general manager of the newly formed Bell Telephone Company.

Theodore Vail began his career as a mail clerk in Omaha for the Union Pacific Railroad and moved to Washington, D.C., to work for the Post Office Department. He became general superintendent of the Railway Mail Service. Hubbard was a member of the Congressional Postal Committee and met Vail. The Vail family had other involvement in telecommunication. Theodore Vail was the second cousin of

Alfred Vail who together with Samuel F. B. Morse gave the first public demonstration of an electromagnetic telegraph. It is thought that Alfred was responsible for the invention of the dot-dash system of code traditionally credited to Morse.

The competitive battle between the Bell company and Western Union soon reached a peak, and in 1878, Bell sued Western Union for patent infringement. At about the same time, Bell received rights to the transmitter invented by Francis Blake, Jr., which was better than Edison's. On February 17, 1879, the National Bell Telephone Company was formed from the other two Bell companies. The patent suit against Western Union was settled on November 10, 1879. Western Union gave up all its patents along with 56,000 telephones in return for 20 percent of Bell rentals for the 17-year life of Bell's patents. Although this might appear as a costly settlement for Bell, all competition was eliminated, thereby giving Bell a monopoly on telephone service. In late 1879, telephone numbers were used for the first time.

With Western Union eliminated as a competitive threat, attention was turned to structuring the telephone industry. On March 19, 1880, the American Bell Telephone Company was incorporated, and Vail became its chief operating officer. At this time, Vail was beginning to form the framework of the Bell System as a three-legged monopoly consisting of *financial control* of the operating companies providing local telephone service; the long lines department of American Bell, which provided *long-distance telephone service*; and *ownership of a manufacturer* providing equipment solely for the Bell System. This framework was a vertically integrated monopoly with strong, central supervision by the parent company, American Bell. With this framework as his objective, Vail then initiated steps to bring it together.

The manufacturer that Vail intended to acquire was the Western Electric Company. Western Electric was formed by Elisha Gray and Enos N. Barton, and it manufactured telegraph and telephone equipment for Western Union. Vail began buying its stock and took controlling interest in November 1881. On February 6, 1882, Western Electric and American Bell signed an agreement in which Bell gave patent rights to Western and Western agreed to supply products to Bell exclusively. American Bell determined the specifications for the telephone equipment that the operating companies were to use, and, of course, Western Electric manufactured its equipment to these specifications. In effect, the operating companies were captive to purchasing all their equipment from Western Electric. Later, in 1908, as a response to antitrust agitation, Western Electric agreed to sell its equipment outside the Bell System also.

The first commercially successful long-distance telephone connection took place on January 12, 1881, between Boston and Providence, Rhode Island, a distance of 45 miles. The provision of long-distance service was to be the glue to hold together the Bell System as a monopoly. Vail realized that his monopoly framework would be difficult to dictate after the basic Bell patents expired. In 1884, he developed the argument that control of the long-distance toll network could be the key to retaining the monopoly after the patents expired. On February 28, 1885, the

American Telephone and Telegraph Company (AT&T) was chartered with Vail as president. AT&T had control over the long-distance lines, and Vail's structuring of the Bell System was now complete. The reason for the word "telegraph" in AT&T would become obvious about 20 years later as AT&T extended its monopoly to include telegraph service too.

In 1887, with his structuring of the Bell System complete, Vail left AT&T. A reason given was his disagreement over the strategy to be followed after expiration of the patents and also over too much emphasis on dividends rather than quality of service. Vail spent the next 20 years in semiretirement sailing, investing, and visiting England. The continued growth of telephone service was phenomenal. In early 1881, there were 133,000 Bell telephones; in 1892, there were 240,000; and in 1899, there were more than 800,000 Bell telephones in service.

Vail's dream of a total monopoly notwithstanding, the independent, non-Bell, telephone companies had not withered away. In 1893, the basic patents expired, and many non-Bell independents entered the telephone business. In 1900, there were roughly 856,000 Bell telephones compared to 600,000 independent telephones. In 1903, the scale tipped, and there were 2,000,000 independent telephones as opposed to 1,278,000 Bell telephones. This had happened because the independents offered better service and lower rates. The concerns that led Vail to leave AT&T were indeed destroying the Bell System. In response to this competition, AT&T dropped rates and also refused to interconnect the independents to the Bell long-distance network. These policies had an effect, and in 1907 there were 3,132,000 Bell telephones to 2,987,000 independent telephones. However, this small victory was obtained at the cost of a poor public image, and Bell still provided poorer service.

The great monopolist, J. P. Morgan, had gained financial control of the Bell System, and in 1902, the Morgan interests placed Vail on AT&T's Board of Directors. On May 1, 1907, Vail again became president of AT&T. Morgan and Vail clearly knew each other well and shared the goal of making AT&T the sole supplier of all telecommunication—telephone and telegraph—service in the United States. Together they bought up the independents. Morgan would cut financial credit to the independents, and Vail would then buy them with a cut-rate deal. The many independents thus acquired were combined with the Bell companies. In 1911, the telephone companies were consolidated and organized essentially as they would remain until divestiture in 1984.

Vail believed that the provision of telephone service was a natural monopoly, but he also recognized that the times were changing and that unbridled monopoly was becoming less acceptable to the public and the government. Therefore, he realized that some form of regulation of the telephone monopoly would be necessary to assure the public and the government that there would be a check on the potential abuse of monopolistic power. State commissions were formed, and welcomed by Vail, to regulate telephone service. By 1915, many states had them, and by 1922,

40 of the 48 had them. These commissions were inadequately staffed and were really little more than rubber stamps of what the telephone companies wanted to do. From all external appearances, however, there was regulation, and the public thought itself protected from the abuses of monopoly.

The Morgan-Vail vision of a monopoly of all telecommunication slowly became reality. In 1909, AT&T acquired 30 percent of Western Union stock, and in 1910 Vail became president of Western Union. Bell's old foe was now part of the Bell empire, and it seemed that Morgan and Vail were about to achieve their goal of a total monopoly, but forces were in motion that would stifle that goal. In 1910, the Interstate Commerce Commission (ICC) was given jurisdiction over the telephone industry, and it began an investigation to determine whether AT&T was attempting to monopolize telecommunication. Furthermore, in 1912, the remaining independent telephone companies protested to the US Justice Department that AT&T was violating the antitrust laws. As a result of these actions, in January 1913, Attorney General George W. Wickersham advised AT&T that certain planned acquisitions were in violation of the Sherman Antitrust Act.

Fate intervened. Morgan died in March 1913, and Vail lost his ally. The AT&T response to the Justice Department came on December 19, 1913, in the form of a letter from AT&T vice-president Nathan C. Kingsbury to then Attorney General James McReynolds. In this letter (known as the *Kingsbury Commitment*), AT&T agreed to dispose of its Western Union stock, to purchase no more independent telephone companies (except with ICC approval), and to interconnect independent telephone companies to the AT&T long-distance network. The Morgan-Vail dream of a total monopoly of all telecommunication was dead. Furthermore, the Kingsbury Commitment set a precedent of compromise by AT&T in the face of government intervention, rather than allowing the government to dictate terms.

In 1918, Bell had 10 million telephones. A year earlier, the United States had entered the World War. Since telecommunication was deemed essential to the war effort, in July 1918, President Wilson acted on the authority previously given him by Congress to assume control of telecommunication companies by placing them under the direction of the Post Office Department. For the following year, the federal government controlled the Bell System. During that time, rates rose sharply, and a service connection charge was instituted. The World War ended in 1918, and on August 1, 1919, the telephone companies were returned to private ownership, thus ending the government's control of the Bell System, much to everyone's relief. In June 1919, Vail resigned as president and became chairman of the board of AT&T. He died about one year later. To this day, he is revered within AT&T and the Bell telephone companies for the leadership and sound management he gave to the Bell System.

The Kingsbury Commitment was, in effect, given legislative approval by the passage of the Graham-Willis Act in 1921 by Congress. The act formally exempted telephony from the Sherman Antitrust Act and allowed AT&T to consolidate independent telephone companies that competed with Bell in the same geographic area.

AT&T's attempts to monopolize telecommunication extended to other industries too. In 1922, AT&T owned its own radio broadcasting station, based on its radio patents. Through the use of the telephone network, a national network of radio stations was created. AT&T impeded competition by refusing to allow rival broadcasters to use the telephone network to create their own competing radio networks. Again, the government was forced to intervene, and in 1925, the government threatened antitrust action against Western Electric. Walter S. Gifford, then president of AT&T, believed that AT&T should emphasize the provision of telephone service rather than these other business activities. Accordingly, in July 1926, AT&T sold its entire radio network to the Radio Corporation of America (RCA) for $1 million, divested its holdings of RCA stock, and agreed to supply wire services to RCA. In other actions, AT&T sold its Western Electric manufacturing plants outside the United States to the recently formed International Telephone and Telegraph Company (ITT). In 1928, AT&T sold its Graybar Electric division, which was responsible for marketing Western Electric products to customers outside the Bell System.

Interestingly, as AT&T was abandoning the radio business, it was at the same time moving into a monopoly position in yet another entertainment business: motion pictures. In 1925, Bell Labs engineers learned how to synchronize sound with moving pictures, and in 1927 Western Electric sound equipment was used to make the *Jazz Singer*. In 1930, 90 percent of all talking motion pictures used Western Electric equipment.

Meanwhile, the telephone business continued to grow and grow. At the end of 1930, there were 15,193,000 Bell telephones in service. In October 1930, AT&T purchased the Teletype Corporation.

The ICC was never that effective in regulating telecommunication, and in 1934 its telecommunication regulatory role was replaced by the creation of the Federal Communications Commission. The FCC almost immediately initiated an investigation of AT&T, but World War II intervened with the Bell System playing an important part in the war effort in such areas as radar, sonar, and telecommunication. Thus, no definitive action occurred in terms of the FCC investigation until January 14, 1949. On that date, charges were filed in the US District Court in Newark, New Jersey, against AT&T for violations of the Sherman Antitrust Act. The case noted the "absence of effective competition" in telecommunication and asked for Western Electric to be divested from AT&T, for the operating companies to purchase their equipment on a competitive basis, and for patents to be licensed to all for reasonable royalties.

The progress of the antitrust case was extremely slow, and the government did not request documents from AT&T until August 1951. Dwight David Eisenhower, a Republican, was elected President in 1952, and the Korean War intervened. Also, AT&T had involved itself in a very strange undertaking for a telecommunication firm, namely, the design and production of nuclear weapons—a task that continues to this day. Sandia Corporation was created in September 1949, at the request of the federal government, to take over the design and production of nuclear weapons previously done under the management of the University of California. The slowness of the government in pursuing the antitrust case might well have been an inducement to get AT&T to agree to managing the development of nuclear weapons.

On January 12, 1956, the antitrust case was settled by a Consent Decree. The Final Judgment, written by AT&T lawyers, stated that Western Electric would remain with AT&T, and the Bell System would remain intact. AT&T abandoned the audio business by having Western Electric sell its audio manufacturer, Westrex. Furthermore, Western Electric agreed to license all its patents and to manufacture equipment only for the Bell System and not to seek other markets. AT&T agreed that the Bell System would engage only in common-carrier communication and other incidental businesses. Thus, through compromise the Bell System remained intact, and the government's attempt to pry Western Electric from the Bell System failed again.

AT&T clearly had won again in its battle with the Sherman Antitrust Act, but a new threat in the form of competitive opportunity created by new technology was about to knock at AT&T's door. In 1955, the FCC allowed *Hush-A-Phone* devices to be attached physically to telephone handsets. The Hush-A-Phone device fitted over the telephone transmitter and prevented one's speech from being picked up. It was simply an acoustic filter. The operating telephone companies refused to allow such devices to be attached to telephones claiming that the devices were "foreign attachments" not allowed by telephone tariffs. The FCC did not agree and allowed the attachments, but in 1956 the US Court of Appeals set aside the FCC decision. This was not the end of the attachment question, however. The next foreign attachment was *Carterphone,* an acoustic coupling device that connected two-way mobile radio conversations with the telephone network. The operating telephone companies refused its use, but in June 1968, Carterphone was allowed by the FCC, and this time the decision stood. As usually happens over time, this crack in the door was widened, and acoustic coupling evolved to direct electrical connection to the network. Consumers purchased their own telephones and connected them directly to telephone lines. The telephone companies tried to prevent this by requiring protective devices to be installed to protect the network. This obstructionistic tactic only delayed the inevitable by a few years. In July 1977, the FCC instituted a certification program whereby any type-certified customer-premises equipment could be connected directly to the telephone network. The Bell end-to-end total monopoly of telephone service was starting to crumble.

Another signal that the days of the Bell monopoly were about to end came in 1962 with the creation by the government of the COMSAT Corporation to handle telecommunication by satellite. AT&T had claimed that satellite communication was simply another form of microwave radio communication and should therefore be controlled by AT&T as just another part of its long-distance network. However, aerospace people saw satellites as part of the space program and an area of competition by private industry. At one time, AT&T actually asked the National Aeronautics and Space Administration (NASA) for exclusive rights to satellite communication, but NASA refused the request. COMSAT was created by the government to be half owned by the communication industry and half by private investors. AT&T was allowed a maximum ownership of 27.5 percent. COMSAT became very profitable. In 1972, the FCC pressured AT&T into selling its COMSAT stock.

In addition to the attack on the telephone terminal market, AT&T's monopoly on long-distance service also came under attack. Microwave Communications Incorporated, now MCI, installed a private-line microwave system between Chicago and Saint Louis. The system did not remain only private line, and it became interconnected with Bell local lines at each end and offered direct competition to AT&T's intercity toll network. After many court and FCC actions, the FCC finally allowed such specialized common carriers in May 1970. Clearly, as AT&T correctly claimed, there was little, if anything, "specialized" about the long-distance services that these competitors offered. But it was equally clear to nearly everyone, if not to AT&T, that the government's policy was to stimulate competition in the telecommunication industry.

AT&T's attempts to stifle competition during these years had not gone unnoticed by the government. Although apparently a surprise to AT&T, on November 20, 1974, the Justice Department filed its antitrust case against AT&T in the US District Court in Washington, DC. The case asked for divestiture of Western Electric and some form of separation between the operating companies and long lines. AT&T, Western Electric, and Bell Labs were named as the defendants, and the operating companies were named as coconspirators.

The case dragged on over the years with AT&T facing great uncertainty about its future at the same time that competition was nibbling away at its basic telecommunication business. AT&T's management had come to believe that AT&T's future survival depended on it having the freedom to enter new high-technology business, such as computers, rather than restricting itself to only common-carrier communications, as agreed in the 1956 Final Judgment. AT&T's historic reaction to fight competitors in the courts and regulatory bodies was consistently failing—it appeared that everyone was now against AT&T. The FCC was insisting that AT&T entries into unregulated business should be kept fully separated from its regulated business. Hence, rather than wait for the antitrust case to come to a final decision, AT&T settled the case "out of court." The historic precedent of compromise was followed once again.

The Justice Department and AT&T reached a compromise agreement in private and then filed the settlement on January 8, 1982, at the District Court in Newark, even though the current case was being heard in Washington. The Newark venue was chosen because the settlement was written as a modification of the 1956 Consent Decree's Final Judgment—thus the term Modified Final Judgment (MFJ). The Final Judgment of 1956 was vacated in its entirety and replaced by the Modification of Final Judgment. The clear tone of the MFJ was that the operating companies were responsible for anticompetitive practices and were to be punished by being divested from AT&T ownership and by being restricted to providing only local exchange service. They were not to be involved in manufacturing or in providing long-distance service—areas that AT&T clearly wished to keep for itself without creating any new competitors. Western Electric remained with AT&T, and they both were rewarded by being freed from all the constraints of the 1956 settlement.

Since the original 1956 case, Congress had passed the Tunney Act, which stipulated that the fairness of antitrust settlements had to be determined by public hearings. AT&T did not want this type of scrutiny nor continued uncertainty and hence attempted to avoid the Tunney Act by making the settlement a modification of the 1956 case. The maneuver failed, and Judge Harold C. Greene angrily had the case returned to his court in Washington.

On August 24, 1982, Judge Greene approved and entered the Modification of Final Judgment. He suggested some changes that were agreed to by AT&T and the Justice Department—although one gets the distinct impression that Justice would have agreed to anything since they too were tired of the case and had actually attempted to drop it during its course. The operating companies were allowed to provide (but not manufacture) customer-premises equipment and to publish yellow-page directories. At the instigation of the newspaper industry, which was convinced that AT&T was about to monopolize electronic publishing, AT&T was forbidden from engaging in electronic publishing over its own transmission facilities for the next seven years, after which time the ban would be re-examined.

Under the stipulations of the Modified Final Judgment, the United States was divided into 160 *local access and transport areas* (LATAs), which were to be the domain of the Bell operating companies (BOCs). The BOCs were forbidden from providing inter-LATA service and were allowed only to provide exchange service within a LATA.

A controversy developed over the use of the Bell name. AT&T claimed ownership, but it was later decided that the Bell name should be reserved for use by the operating companies. Bell Labs was the only AT&T entity that was allowed to continue to use the Bell name, but had to indicate the AT&T affiliation by calling itself AT&T Bell Labs.

Before the actual divestiture occurred, AT&T reorganized the operating companies into seven *regional holding companies* (RHCs) and then simply divested the stock of the RHCs. Prior to divestiture, AT&T also created a central organization to perform common engineering and applied research for the regional companies—this organization became Bell Communications Research, or Bellcore.

Thus, AT&T's monopoly of telecommunication in the United States and the Bell System finally came to an end. The actual divestiture of the operating companies occurred on January 1, 1984. The operating companies are now all independent of AT&T, as some of them were before being acquired by Vail nearly a century ago.

REFERENCES

Brooks, John, *Telephone: The First Hundred Years,* New York: Harper & Row, Publishers, 1976.

Goulden, Joseph C., *Monopoly,* New York: Pocket Books, 1970.

Kinsley, Michael, *Outer Space and Inner Sanctums,* New York: John Wiley and Sons, 1976.

von Auw, Alvin, *Heritage and Destiny,* New York: Praeger, 1983.

Chapter 9
COMMUNICATION SERVICES: OLD, NEW, AND DOUBTFUL

INTRODUCTION

The technology of today's telecommunication network is very impressive and sophisticated, but it is not technology *per se* that consumers purchase. Consumers purchase services and what the technology can do for them. This chapter describes and discusses telecommunication services along with the acceptance and rejection of these services by consumers. Some of these services are old and are well accepted by consumers. Other services, such as cellular telephone service, are brand new, but are achieving rapid acceptance in the marketplace. Still other services have had a doubtful future and perhaps will continue to have a poor future.

The telephone has been such a great success that we sometimes forget that telecommunication has not been without its failures too. These failures are not failures of technology, but rather are failures of markets to appear for some of the new products and services made possible by the technology. In many cases, the market failures occurred because the consumer had little or no use for the products and services. Furthermore, consumer needs and behavior in using novel forms of communication were frequently poorly understood.

The following sections of this chapter discuss market laggards and outright failures, all involving forms of visual communication. The first is two-way interpersonal visual communication: the video telephone, or picturephone, as it was called. The next is a form of group teleconferencing. The last is the remote access of text and graphic information for the home: videotex. I describe these technologies so that we can learn about the nature of human communication, see the future more clearly, and be more cautious in our enthusiasm for technology.

Later in this chapter, I describe services that have succeeded. We shall see that the pulic switched network is a great resource making possible a host of useful services—services that enable us to summon emergency assistance, to make toll-free calls, to transmit documents, and to reach people wherever they may be. In many respects, the emphasis in this chapter on services, including those that have succeeded along with those that have failed or will probably fail, creates a glimpse of what I believe is the future of telecommunication.

LESSONS FROM PICTUREPHONE

It would seem that every industry has its "Edsel"—some technology, product, or service that has failed in the marketplace, despite extensive efforts to promote and sell it. The telecommunication industry's Edsel is the picturephone—a telephone that enables you not only to hear but also to see the other party. In return, you are not only heard by the other party but are seen too.

In the late 1920s, Bell Laboratories conducted research into a visual interpersonal communication system; it would later evolve into television. AT&T demonstrated the results of this research at the Chicago World's Fair in the early 1930s as a two-way, color, video telephone. In the late 1950s, Bell Labs initiated development of a trial videophone. The results of this development effort were demonstrated at the New York World's Fair in 1964 as the picturephone system. AT&T interpreted the consumer repsonse to the demonstration as strong indication of a positive market need, although 700 interviews were conducted only of people who used the demonstration picturephone system, and even then, only one-half of them said it was important to see the other person during a telephone conversation. Thus, in the mid-1960s, AT&T made the decision to introduce picturephone service in 1970. Indeed, after much development work, picturephone service was introduced in 1970 in Pittsburgh, and at that time AT&T forecast 41,500 picturephone sets in use in 25 cities by 1975. Picturephone service was introduced to Chicago in 1971, and at that time the Institute for the Future performed a Delphi study that forecast over 2 million picturephone sets in 1985.

The picturephone system introduced in the early 1970s worked over conventional twisted pairs of copper wire on local loops of less than six miles in length. Three twisted pairs were used: one for two-way audio and two for two-way video. The video bandwidth was 1 MHz—much less than over-the-air television's 4.5 MHz. The bandwidth reduction was achieved by lower resolution: 250 lines per frame versus 525 lines for broadcast television. Like broadcast television, 30 full frames were transmitted per second. The picture size was a small 5.5 by 5 inches designed to show only the head and face of the user. Picturephone service cost a flat $160 per month for the first 30 minutes and then $0.25 per additional minute.

The market acceptance of picturephone service was very poor, and in 1972 AT&T cut back on the manufacture of picturephone sets, after having spent an estimated $500 million on the project. Amazingly, AT&T had developed the picturephone system and had introduced it into the marketplace with only the most meager evidence of any real market need—it was mostly hope and belief that guided the picturephone project.

With the poor market acceptance of its picturephone service, AT&T finally initiated extensive consumer research. A study conducted in 1971 showed that strict face-to-face applications of picturephone communication had limited appeal and that "lack of application" was the major reason for not subscribing to the service. However, this study did show that document display had a relatively wide interest. In 1972, AT&T admitted its problems with the picturephone system and announced that it was about to begin an exploration program to study the market for two-way, switched visual communication. A picturephone system had been installed for use by AT&T management, and the results of a study of these users conducted in 1974 indicated it would make little or no difference to them if picturephone service was discontinued. Two-thirds of picturephone users used it for applications involving graphics, and consumer studies conducted in 1974 indicated that the greatest need was for high-powered graphic communication and that there was "little need for face-to-face communication" over a picturephone system. A study conducted in 1975 of past executive users of picturephone service concluded that it was "a bust." The final conclusion of all these studies was that picturephone service was "a concept looking for a market" and "an idea ahead of its time." With hindsight, we see that the continued expression of need for some form of graphical communication predicted the recent success of facsimile.

As part of its market exploration program, AT&T searched for high-need applications for two-way, switched visual communication. One such application was the picturephone system that was operated at the Bethany-Garfield Hospital in Chicago from March 1973 to April 1976. About 20 picturephone sets formed the system, and usage averaged about 1800 calls per month. The US Department of Health, Education, and Welfare funded the costs of the system as part of its research into new telecommunication technologies. Although the system was being used, I was suspicious of why, particularly when I personally visited the installation and noticed that the sets were often so positioned that the person at the other end received a picture of the user's belt buckle. Hence, my colleague at AT&T, James P. Woods, and I conducted a study to determine why people were using the system.

Interviews were conducted of the 57 hospital people who had access to the system. The results showed that the system was used primarily because of its ability to reach people quickly. Because of its distinctive ring and because it could be used only for internal calls, the picturephone was, in effect, a "hot line." Another reason for its use was the ability to show forms. The ability to see the other person was a distant third in the order of reasons for use. Only the people who used the picturephone system the least rated the ability to see the other person as its best feature. When the federal funding ceased, the hospital had the system removed.

A picturephone system was installed in Phoenix in the 1970s for use in the criminal justice system so that attorneys could use the picturephone to see and to talk to their incarcerated clients rather than physically going to the jail. The visual demeanor of a client was important to assess for an attorney, and thus the use of the picturephone system seemed to make good sense. However, the same type of service could have been provided by a simple closed-circuit, two-way television system. The flexibility of the switched picturephone system simply was not needed in this application.

The virtually unlimited bandwidth of optical fiber, coupled with digital switching and increasingly less costly television cameras and displays, creates a hope that the environment for picturephone service is becoming more positive. This hope assumes that economics was a major cause of the failure of picturephone and that newer technology is the solution. I think not. Picturephone failed for the simple reason that it was not an improvement over a conventional telephone call. The intimacy of a telephone call is shattered by seeing a face on a screen. One must sit still to remain on camera and must be concerned with visual impression. Based on informal study, I am convinced that many people would actually pay extra *not* to have picturephone service if it were standard. A few years ago, small two-way visual telephones that worked over regular telephone circuits were introduced. These visual telephones used algorithms to compress and reduce the bandwidth needed to transmit the visual image. The resolution of the image was poor, although perhaps just acceptable for seeing a face. These visual telephones failed in the market, just like AT&T's picturephone, confirming my negative view of the consumer need for such a form of visual communication. Yet, some of my friends and colleagues, although they may otherwise appear quite sane, continue to believe that the picturephone will ultimately replace the telephone.

TELECONFERENCING

The energy crisis occurred at the time of the visual communication market exploration program, and this led to an investigation of two-way video teleconferencing by AT&T. AT&T's videoconferencing system interconnected such major cities as New York, Washington, San Francisco, Los Angeles, and Chicago. The service was available at public rooms located in AT&T buildings in these cities, and the rooms in any two cities could be connected. In response to customer comments, continual improvements were made in the technology until it finally offered color video at the NTSC broadcast standard, overhead graphics, and full-duplex audio. The rooms were designed to be pleasing and easy to use. Full-time attendants staffed the rooms and were available to assist the users. The rates were set assuming future techological advances to reduce bandwidth and were only about $400 an hour for a coast-to-coast teleconference.

The teleconferencing service was demonstrated to many AT&T customers, and nearly all who saw the system were impressed and were positive about using it. Some actually came and used it for a meeting or two, but the problem was that few, if any, came back to use it again. I remember a focus-group session in which the participants were all positive about the service based on a demonstration, but had not used the system again. At the end of the session, the moderator stated that AT&T would offer a free call to the participants—not one participant was interested! It was then that I realized we could not even give video teleconferencing away for free! This poor response was replicated with an internal videoconferencing system that was installed in July 1975 between an AT&T building in Morristown, New Jersey, and AT&T's corporate headquarters in New York. A study predicted heavy daily usage of the system, but after six months, actual usage was quite low.

Clearly, the problem with video teleconferencing was not technology or economics. The answer came from research conducted in England and in the United States into the purpose of meetings. The results of this research indicated that teleconferencing was most useful for meetings characterized by the exchange of information between participants who knew each other. Two studies that I conducted indicated that such "target market" meetings are only 4 percent of the total of all types of group meetings. Thus, AT&T's difficulty in promoting video teleconferencing was involved with the types and purposes of meetings. We also discovered that few people want to use public rooms and facilities.

I personally believe there is a market for teleconferencing of all kinds, from the simplest audio-only systems to elaborate full-motion video systems. One of the most impressive video systems I have ever seen is Bellcore's VideoWindow® system, shown in Figure 9.1. I advocate a "product line" approach to teleconferencing in which a spectrum of technologies is offered depending on distance, budget, and types of meetings. For many applications, simple audio-only systems augmented with facsimile are sufficient. Full-motion video is the "Cadillac" of teleconferencing, and, even with modern bandwidth compression, is still quite costly.

Claims have been made that in 1990 there were more than 2500 public and private video teleconferening rooms in the United States. Perhaps we should be impressed with this demonstrated "success" of video teleconferencing, but when I think of the vast numbers of public and private telephones, facsimile units, and cellular telephones, I do not get excited about video teleconferencing. The "problem" with teleconferencing of all kinds is that hyperbole and hope far outstrip realistic expectations. It fascinates me that such an overwhelming enthusiasm can still exist toward video teleconferencing, given its history of low market acceptance and use, yet in 1990 AT&T announced with great enthusiasm the availability of a two-way video teleconferencing service between public rooms in New York and Moscow at the rate of $3200 per hour. I suspect that it will be idle most of the time.

VIDEOTEX

IBM and Sears, Roebuck and Company reportedly spent over $500 million in developing their *Prodigy*[SM] service, which was introduced in 1988. As of 1990, there

Figure 9.1 VideoWindow® is a video teleconferencing system developed by Bellcore engineers. Two video images are projected side-by-side using rear-screen projectors and are life size. There is no switching of cameras, and it appears as if the remote participants are sitting there on the opposite side of the room. High-quality, full-duplex audio is also used facilitating normal conversation. The system is installed between Bellcore facilities in Morristown and Nave-sink, New Jersey, a distance of about 70 miles by car. (Photo courtesy of Bellcore. © 1990 Bellcore.)

are claims of about 500,000 subscribers in the United States. Prodigy service is a computerized database containing all sorts of information that can be accessed over the telephone network using a personal computer and a modem. In France, a small Minitel terminal, which also can be used as a telephone, enables access to databases containing telephone listings and other information. These types of database services are known as *videotex.*

Videotex is not new. The British Post Office spent over $50 million during the early to mid-1970s developing its Prestel videotex service, which was introduced in 1979. The Prestel service was designed to use the home TV set for display of text and graphic information retrieved over telephone lines from the central database.

The information was organized in a tree-like fashion to make it easy to find and retrieve a specific page of data. Sports, weather, shopping, finance, and travel information were some of the information services that were available. The British coined the term *viewdata* to apply to their service. The acceptance of the service by British consumers was considerably less than expected.

The US market for videotex was explored in the early 1980s. AT&T joined with Knight-Ridder Newspapers in a videotex service, called *Viewtron* service, which was initially offered on a trial basis to participants in Coral Gables, Florida. AT&T supplied the terminals to attach to home TV sets; Southern Bell supplied the network; and Knight-Ridder formed a subsidiary, the Viewdata Corporation of America, to supply the database. The service that was used the most was a text-based messaging service, known as a bulletin board, in which the participants could enter and store messages for each other. AT&T and Knight-Ridder became enthused about the Viewtron service and decided to launch it commercially. However, consumer acceptance was far less than expected, and the service was withdrawn. At about the same time, AT&T joined with Times-Mirror to develop and offer a similar videotex service in Orange County, California. I remember visiting the Times-Mirror videotex office and attempting to use the service to determine whether there was a certain department store in the mall across the street. After spending an hour searching the database, we gave up and used the telephone book. The Times-Mirror videotex service met with low consumer acceptance, and it too was terminated. Similar market failures of videotex occurred around the world in such countries as Canada, Germany, and Sweden.

Disappointing consumer acceptance appears to characterize videotex. Only the French Teletel service is considered a "success," and that is because the French government gives away the Minitel terminals. The Minitel is an integrated telephone and simple black-and-white display terminal. It apparently is easy to use, and the French have created a widespread packet-switched data network to support the system. Millions of Minitel terminals are in use, and the French are actively promoting their system around the world. The largest use of the Teletel service is to obtain telephone numbers from the database, and the second largest use is to send text-based messages. In 1988, there was less than one Minitel call per day—much less than telephone use in the United States where ten telephone calls per day are the norm. Also, most of the present use of Minitel is by businesses, which means that home use is probably much less than one call per day. Are Minitel terminals gathering dust in many French homes? Since videotex in many different implementations has failed consistently around the world, is France somehow a fluke or have the French found the equation that makes videotex work? Do the French have a unique fondness for the written word and for sending text messages? A study done by the French government accounting office indicated that Teletel was quite costly and had accumulated quite a large loss, despite claims to the contrary by the Teletel people.

The IBM-Sears Prodigy service cost so much to develop and costs so much to operate and market that, like the French Teletel service, it too is deeply in the red. It would require tens of millions of users to turn a profit and recoup the initial investment. Given the history of failure of videotex, it is clear that the ambitions of IBM and Sears are far from realistic.

All this is not to say that there is *no* market for videotex—there simply seems to be no *mass consumer* market for videotex. Computerized database services of all kinds and varieties abound in the United States. Both the Source and CompuServe offer computer hobbyists and others a primarily text-based videotex service used for meassaging and access to text-based databases. Roughly the same number of people subscribe to these two services as to the Prodigy service, and I would guess that there are many duplicate subscribers to all three. There also are a number of database services used by professionals, such as the LEXIS® service offered by Mead Data Central to legal professionals and the MEDLINE system operated by the National Library of Medicine for physicicans and medical researchers.

The largest on-line databases are those operated by the airlines to make airline reservations. The SABRE® system is the largest real-time computer network and is operated by American Airlines. The second largest airline reservation system is the Appollo® system provided by United Airlines. Each of these systems is capable of handling over 1500 entries per second and each serves over 10,000 travel agency locations around the world. On a typical day, each system handles about 400,000 reservations.

At one time I believed in videotex and encouraged AT&T to investigate this market. I suggested that one approach for AT&T to enter this market would be to offer an electronic version of the telephone directory, including its classified section, that could be accessed by a simple terminal. I also believed that people had a need to send text messages to each other—a form of electronic mail. As a strategy to enter that market, I suggested the development of a text terminal that was compatible with the terminals used by the deaf.

I guess I now see the futility of all this. I am not going to turn on my personal computer, load software, and connect my modem simply to obtain a telephone number. It is far easier to call directory assistance. But if my telephone were also a terminal, like the French Minitel, I might be more willing to use it to access a remote database of telephone numbers and classified information. Electronic mail is available today, but the many systems are all different and are too cumbersome for me to use—facsimile is much simpler and easier. I most certainly have no need for access to some videotex database that purports to contain all the information I would ever want—I simply do not have the time or patience for such nonsense—and the daily newspaper does it better. Let the computer hobbyists have their toy; it is not for me.

But it would be nice to know tomorrow's weather or today's traffic conditions without waiting for the radio announcer to finally tell me what I want to know. It would be nice somehow to have instant access to such general, timely information. Although we do not have it in the United States, access to such information is possible in Europe, and particularly Scandinavia, through a system called *teletext.* It is a success!

TELETEXT

Teletext was invented in the 1970s by the British and transmits a few hundred frames of text and graphic information over the air along with the standard television signal. The frames of information are inserted in the vertical retrace interval of the TV signal and are retrieved and displayed on the home TV set using special decoding circuitry in the set. The frames are sent in a round-robin fashion requiring about 30 seconds to send a complete set of frames. On the average, about 15 seconds is required to retrieve a specific frame.

Teletext is a success in Europe. In England, nearly every TV set sold there is teletext equipped, and more than one-third of TV households have the service. Teletext is free. If it were available in the United States on a widespread basis, one could envision a scheme in which it would be financed by giving local information in support of national commercials. Teletext is easy to use, is free, and offers fast access to timely, general information of a mass-market appeal, such as weather reports, traffic conditions, TV listings, and news headlines. The TV remote control is used to specify the number of a desired frame of information. Confusion over standards, questions about how to finance its operation, and a fascination with high-definition TV appear to be some of the reasons teletext is not available or promoted in the United States. The success of teletext around the world is another example that the United States is no longer at the leading edge of consumer electronics.

ELECTRONIC HOME BANKING

Electronic teller machines are a great success. Consumers have a need to obtain funds at all sort of hours, and it is normal to see lines of people waiting at a teller machine late at night to obtain cash for some purpose or another. The extension of banking to the home through the use of a terminal at the home would seem to make sense, and a number of banks have conducted trials of home banking, frequently in concert with some videotex system. The videotex terminal displays balances and allows the user to move funds from one account to another and to pay bills. Unfortunately, it does not dispense cash!

For me, it is too simple to write a check and insert it in an envelope to pay a bill. I do not have the time to use a terminal to type electronic checks, nor would I trust such a system without keeping my own written records for verification purposes. My need is an easy way to get cash, and here the home terminal is no help. I guess that my behavior was representative of most other consumers, and most attempts at home banking with some videotex or video terminal of some kind have met with failure and have been cancelled.

However, I would like to be able to determine whether some expected deposit has arrived at the bank, but telephoning the bank to speak to an employee is too much of an effort, except in the most serious cases. My bank loses deposits and makes other mistakes, and hence I have a need to verify deposits and balances. An answer is the use of *audiotex*.

AUDIOTEX

With *audiotex,* a touch-tone phone is used to enter digits, such as the bank account number, and to request various services, such as the account balance. The bank's computer responds by reading back the requested information using recorded or synthetic speech. Audiotex (sometimes spelled audiotext) is easy to use, and no sophisticated training or terminal is required. Audiotex banking has been a great success, one example being Citibank's CitiTouch[SM] system.

I remember a few years ago, when I returned to campus in September, that the usual long lines of students waiting to register for class were not there. My first thought was that the university had lost all its students and I would soon be without a job! Audiotex was the cause. Students now use audiotex to register for class from home or a dormitory room. At least one airline offers an audiotex system so that people can check on expected arrival and departure times of their flights. The system, like most audiotex systems, is quick and easy to use and saves labor costs for the supplier. Audiotex is a real winner, in my judgment, and I expect to see its use continue to grow. Its success is based on its use of the existing telephone network.

FACSIMILE

The invention of *facsimile,* or *fax* for short, in the 1840s predates the telephone. In 1977, only about 130,000 facsimile terminals were in use in the United States; today the number in use is many times that number. Facsimile exploded in use over the past few years. One reason for this success is the ease of use of facsimile; another reason is the familiarity and utility of a paper copy. Facsimile works over the switched public telephone network and thus goes everywhere there is a telephone line. A modern facsimile machine is shown in Figure 9.2. At one time, one vendor's facsimile machine would not work with another vendor's machine. This compatibility problem was solved by the introduction of standards by the Inter-

Figure 9.2 Photograph of a modern facsimile unit. Such units incorporate many features, such as unattended reception and automatic dialing. (Photo of Omnifax® facsimile machine courtesy of the Telautograph Corporation.)

national Telegraph and Telephone Consultative Committee (CCITT), and this was a big factor in facsimile's great success.

A facsimile system consists of a scanner that scans and converts a document for transmission and a recorder that reproduces the document at the receiving destination. Usually, a scanner and a recorder are located within a single facsimile machine so that it can be used both for transmitting and for receiving documents; such a machine functions as a *transceiver.* The earliest scanner consisted of a drum around which the document was wrapped. The drum rotated while a screw mechanism moved a photocell along the length of the rotating document. The recorder was similar except that a writing element was used with some form of photosensitive or electrosensitive paper. The process was slow: six minutes were needed to send a single page.

A modern fax machine uses a linear array of photodiodes to scan a whole line at a time. Typically, 1728 diodes are in the array. An optical system images the line to be scanned onto the diode array. The document is moved along a line at a time until the scanning process has been completed. The output signal from each diode represents the information in a small picture element on the page. An alternative method of scanning is the use of a laser beam: the amount of laser light reflected from the page as it is scanned is sensed. These types of scanners are called flat-bed scanners, as opposed to the older drum scanner. If represented as only black or white, a scanned picture element is called a *pel;* if a gray scale is represented, the scanned picture element is called a *pixel*.

A number of methods are used for the recorder in a modern fax machine. Plain paper is used with a xerographic process. A laser beam writes the information onto a photoreceptive drum, toner is attracted to the charged portions of the drum, and the toner is transferred to the paper and fused to the paper by heat. Heat-sensitive paper is used with the thermal recorder. A write head with 1728 fine wires, corresponding to the diodes in the diode array, heats the paper as it passes through the recorder.

The CCITT established three groups of standards for facsimile machines that work over the existing telephone network. A fourth group is intended for use on end-to-end digital networks of the future. The Group 1 standard transmits an 8.5- by 11-inch page in six minutes at a resolution of 96 scan lines per inch. A sine-wave carrier is shifted in frequency to indicate white or black. The Group 2 standard transmits a page in three minutes at a resolution of about 100 scan lines per inch. Vestigial amplitude modulation and phase modulation of a 2100-Hz carrier is used. This more efficient method of modulation allows twice as much information to be transmitted compared to the Group 1 standard.

The Group 3 standard transmits a page in about one minute at a resolution of about 100 lines per inch. This decrease in transmission time is obtained through the use of run-length encoding of each scan line. With run-length encoding, the length of a run of white or of black picture elements on a scan line is encoded. Typical documents were studied to determine average run lengths, and a modified Hoffmann code was then used to encode the run lengths to ensure a minimum average-length code word. Each scan line has a resolution of 1728 picture elements. Since the encoding scheme examines only a single scan line at a time, it is called one-dimensional coding. A two-dimensional coding scheme that examines two scan lines at a time is available as an option. This scheme is called the *relative element address designate* (READ) code. It encodes one line relative to the preceding line.

Group 3 machines contain a modem and transmit digital information over the telephone network. Rates as high as 9600 bps can be used on good connections, but the rate decreases automatically on poorer connections to as low as 2400 bps on the poorest connection. A higher resolution of about 200 lines per inch is available for Group 3 machines, but more transmission time is needed.

Facsimile has become quite ubiquitous. Many sandwich shops in large cities have fax machines to receive take-out orders from their business customers. Because of time zone differences, fax is the only way to communicate quickly with people on the other side of the Atlantic. Hardly a week goes by that someone does not ask for my fax number. Facsimile has become so popular that the receipt of junk faxes has become a problem.

TELEWRITERS

An early predecessor of the facsimile was the telautograph machine, invented by Elisha Gray in the mid-1800s. The telautograph machine enabled the user to transmit handwriting over telecommunication facilities to a distant location, in effect, the telecommunication of writing, or *telewriting*. Today's version of his invention consists of a conventional pen at the transmitter terminal. As the user writes on a special tablet, the movement of the pen is sensed and is transmitted to the receiver where a second pen transcribes the identical movement on a piece of paper. The system is very simple, yet effective.

Years ago, I installed a telewriter system between my secretary's desk and my office. My secretary picked up my telephone when I was out of the office or when I was speaking on my line. By using the telewriter system, the secretary could write and send me messages that would be recorded instantly on paper on the telewriter receiver in my office. If I was using my telephone and another call came in, I could know immediately who called and the message. This was more efficient than the hand delivery of message slips.

Another application of telewriters is from the sales floor to the stockroom, as shown in Figure 9.3. The salesperson writes the order on a telewriter machine connected to the stockroom. The clerk in the stockroom then retrieves the item and delivers it to the sales floor, perhaps in a small elevator. I saw such a telewriter system in use in a jewelry store on Fifth Avenue in New York City a few years ago.

I would imagine that small keyboards and computer terminals will ultimately displace many of the applications for telewriters, and, as shown in Figure 9.4, the Teleautograph Corporation already makes such a small keyboard unit. But meanwhile, in this day of overhyped technology it is interesting to see a 100-year-old technology still in use!

KEEPING IN CONTACT

The airplane version of cellular telephony is GTE's Airfone® service. Nearly a third of the US commerical fleet is equipped with an Airfone terminal. The user deposits a credit card in the terminal and removes a portable handset. A conventional telephone call is then made using the handset. The airborne terminal communicates with a network of about 80 ground stations that cover the continental United States. A total of 310 two-way radio channels are used in the radio-frequency band from 850 to 899 MHz. Each ground station uses 31 of these channels. The ground stations connect to GTE's long-distance network. Although calls can be made from the Airfone terminal, calls from the ground cannot be made to the airplane. Thus, you are still safe from being reached once you are in the airplane!

Figure 9.3 Photograph of a telewriting system in use in a shipping department. (Photo courtesy of the Telautograph Corporation.)

Nearly 10 million people in the United States subscribe to paging services. We are all familiar with the conventional pager that is worn on one's belt. Newer pagers are so small that they can be worn on the wrist like a watch, as shown in Figure 9.5. Older pagers simply beeped, and the user went to the nearest phone to call the paging office. Newer pagers have small liquid crystal displays of the telephone number of the party to be called back. In some systems, the radio signal sent to the pager is carried on a subcarrier of a local FM radio station; in other systems, the radio signal is sent at a frequency of about 1 GHz. In some paging systems,

Figure 9.4 Photograph of an Omninote® terminal that is used for sending short messages from one office to another. The existing electrical wiring in the building is used to carry the signals between the terminals. (Photo courtesy of the Telautograph Corporation.)

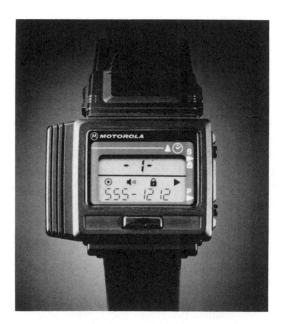

Figure 9.5 Photograph of a wristwatch pager. (Photo courtesy of Motorola, Inc.)

communication satellites carry paging requests across the continent to metropolitan areas where local paging signals are then sent by radio.

800 SERVICE

I find it unbelievable but about one-third of the business calls carried on AT&T's long-distance network are toll-free 800 calls! Though free to the caller, the called party does pay for the call. I use 800 numbers to order goods from catalog stores all over the country. Salespeople on the road use 800 numbers to call their offices to place orders. Today's 800 service is very sophisticated and offers such features as the ability to transfer incoming calls to different offices across the country depending on the time of day or the traffic load.

800 service offered by AT&T has a database located centrally in the network. The 800 number is forwarded to that database and is converted to the appropriate routing to reach the called party. The routing information is sent over the common channel signaling network, and the voice connection is finally made.

EMERGENCY SERVICES

The 911 service is a real lifesaver. The newest version is called Enhanced 911 service, or E911. Calls are automatically routed to the appropriate Public Safety Answering Point (PSAP). With the older system, calls might be sent to the wrong PSAP if the boundaries of the telephone exchange did not agree with municipal boundaries. E911 includes a centrally located database of addresses that is automatically accessed when a 911 call occurs. The address information is sent to the PSAP where it is displayed on a terminal so that the dispatcher knows where to send the emergency response unit. This feature is known as *automatic location information* (ALI) and is particularly useful if a small child calls or if the caller does not have time to give the address. With E911 service, the telephone number of the calling party is automatically sent to the PSAP and displayed there, a feature called *automatic number identification* (ANI). In this way, the dispatcher can always call back the emergency caller if they are disconnected for some reason.

In many states, 911 service is financed by a small surcharge added to the monthly telephone bill. This most certainly is a small price to pay to save a life, but all the cents do add up to millions of dollars. The use of common channel signaling (SS7) in the local network should decrease the cost of providing 911 service in the future.

About one-quarter million people in the United States are protected by the Lifeline® emergency response service provided by Lifeline Systems. This service, and similar systems operated by others, uses a small device carried or worn by the user in the home. If an emergency occurs, the user pushes a button on the device, and a radio signal is sent to a receiver connected to the telephone line in the home. The receiver then automatically places a telephone call to a central location and transmits the identity of the emergency caller. The central location then dispatches an ambulance or other appropriate response.

TELEVISION

Although I have my doubts about the market viability of two-way video communication—the picturephone—there is no doubt about the viability of one-way broadcast video communication—television. It was a great commercial success right from its earliest days. I remember when neighbors would all congregate at the home of the one family who had the only television set in the neighborhood. How will television change in the future? Will television broadcast be digital someday? Will digital television signals be sent over the air and over CATV systems? This will take a lot of bandwidth, but with optical fiber, the bandwidth is there. Also, image processing by computers could reduce the bandwidth to reasonable amounts.

In 1989, CATV was available to over 85 percent of TV households in the United States, although only about 55 percent actually subscribed. As the remainder of the country is wired with coax, and perhaps with fiber, nearly all will have the potential of receiving television over a land-based system. Then, will television broadcast over the air be obsolete? As national policy, would it make sense to require CATV firms to offer basic broadcast television to everyone for free and to charge only for nonbasic services and what should be defined as "basic"?

If television is delivered to the home over a broadband medium, such as optical fiber, will the consumer want all the new services that will be possible? One such service is video-on-demand—the ability to request any TV program or video for viewing at any time. We do this now using our VCR and going to the video store to rent a tape. With video-on-demand we can stay at home and "dial up" any video. But a tape rents for about $1 per hour; fiber to the home will be expensive, and will we be willing to pay what it might cost to see a video over such a costly medium? Maybe someday the cost will be low enough.

Another new TV service is high-definition television. HDTV offers twice the vertical and horizontal resolution of present television along with a wider screen, but it requires a whole new TV standard along with a lot of bandwidth. Do you watch TV because of its image quality or because of program content and quality? Will we be willing to pay a premium of $500 or more for a HDTV set? Will the TV stations be able to afford the costs of replacing all their cameras, recorders, monitors, and other equipment with new HDTV equipment? If the TV set is a conventional screen size and is viewed at its usual distance, will we notice the improvement of HDTV? No, according to research performed thus far. Conventional 525-line TV was designed with the resolution of the human eye in mind and an assumed viewing distance of no closer than four times the screen height. Increasing the number of lines will not improve things unless you sit closer to the screen or the screen is much larger. Thus, some people believe that the market for HDTV is a large screen on the wall and the bringing of wide-screen movies into the home—but do you have a whole wall to allocate for use as a TV screen? I do not.

Perhaps all this excitement about HDTV is fluff. But someday TV will be digital and some form of improved TV or HDTV will be available. Perhaps we will have small-screen, portable, personal HDTV sets that play videotapes along with programs received over the air. Someday, but when!

THE BROADBANDWAGON

Some years ago, in the 1970s, it was thought that the coaxial cable medium used by the CATV firms could be made two way and could hence carry a host of two-way services, such as voice telephony and data, in addition to the one-way television signals. Video-on-demand in which the user requests some specific video was also suggested. Some trial systems were installed, but in the end, most of them failed. The coaxial cable of CATV is a one-way broadcast medium used primarily for entertainment purposes.

Optical fiber to the home is a broadband medium that today is being promoted for delivering a host of two-way services, such as voice telephony and data, along with one-way television signals. All this sounds quite familiar to the past promotion of coaxial cable. However, coaxial cable is controlled by the CATV firms, and optical fiber thus far is expected to be controlled by the telephone companies. In the past, it was the CATV firms that were looking for services other than the delivery of television signals. Today, it is the telephone companies that are looking for the delivery of services beyond telephone service. A number of policy issues cloud whether telephone companies will be allowed to control the conduit to the home over which television shows will be delivered. The issue of a single broadband medium to the home has pitted the telephone companies against the CATV industry.

From an engineering perspective, the use of a single broadband medium to carry a number of services looks intriguing. However, entertainment video signals are mostly one way and are broadcast to everyone. Communication signals, such as voice telephony and computer data, are two way and are sent from one individual to another. The two worlds are quite different. Perhaps the ultimate solution is the use of two broadband media to the home: one carrying one-way television signals and the other carrying a number of two-way communication signals. Meanwhile, now is the time to get on board the broadbandwagon before you are left behind! We shall see whether it turns out to be another picturephone (a failure) or facsimile (a success).

REFERENCES

Costigan, D.M., *Electronic Delivery of Documents,* New York: Van Nostrand Reinhold, 1978.

Hunter, Roy, and A. Harry Robinson, "International Digital Facsimile Coding Standards," *Proc. IEEE,* Vol. 68, No. 7, pp. 854–867.

Noll, A. Michael, "Teleconferencing Target Market," *Information Management Rev.,* Vol. 2, No. 2 (Fall 1986), pp. 65–73.

Noll, A. Michael, "The Broadbandwagon," *Telecommunications Policy,* Vol. 13, No. 3, September 1989, pp. 197–201.

Noll, A. Michael, "Videotex: Anatomy of a Failure," *Information & Management,* Vol. 9, 1985, pp. 99–109.

Noll, A. Michael, and James P. Woods, Jr., "The Use of Picturephone Service in a Hospital," *Telecommunications Policy,* Vol. 3, No. 1, March 1979, pp. 29–36.

Chapter 10
PERSPECTIVES

INTRODUCTION

The Bell System had grown during its century of existence until, at its zenith in the early 1980s, it had about one million employees, profits of over $7 billion, revenues of about $70 billion per year, and total assets of nearly $150 billion. In 1984, this corporate empire was toppled. The story of how that occurred along with some personal perspectives on that story are the subject of this chapter. I also cannot let our story of telephone communication end without giving some final thoughts and perspectives on its future.

The Bell System was invented over one hundred years ago by Theodore N. Vail, who was president of the American Telephone and Telegraph Company from 1907 to 1919. Telecommunication was a hodgepodge of competing companies during the early days of telephony, particularly following the expiration of Alexander Graham Bell's patents. Vail believed that all this competition did not make much sense and rationalized this belief by claiming that the public would be served best if telephony service was provided by a single company on a nationwide scale. The concept of a natural monopoly for telephone service was thus promulgated, but Vail's true genius lay in his invention and endorsement of public regulation of the monopoly to ensure that the public interest was best served.

Both Bell and Vail believed in total end-to-end provision of telephone service, including the telephone instruments, the wiring in homes and businesses, and the transmission and switching equipment and systems that interconnected the telephones. Bell and Vail had the vision to realize that profits would ultimately be greatest in the long term if everyone could afford to have telephone service. Thus, the concept of universal service for all was born.

The goal of universally affordable telephone service was achieved years ago. This goal was achieved partially by using such services as long distance and touch-tone to subsidize local service. This subsidizing later created opportunities for competition in the sale of telephones and in the provision of long-distance telecommunication. The insistence by the Bell System on total end-to-end control of telephone service became less and less tenable. Consumers could purchase and connect their own telephones to the local line. A number of firms were supplying long-distance service in open competition with AT&T. The Bell monopoly was crumbling, and its final official fragmentation occurred in 1982 under the terms of the Modified Final Judgment, reached between AT&T and the US Department of Justice.

This chapter explores the causes and ramifications of this fragmentation of the Bell System. Technological, historical, financial, and my personal views are integrated in this chapter to give the reader some perspectives on the past, present, and future of telecommunication.

BELL SYSTEM ORGANIZATIONAL STRUCTURE

The organizational structure of the Bell System was determined nearly three-quarters of a century ago and persisted until the restructuring that was completed in early 1984.

The Bell System consisted of four major units operating as a single, vertically integrated monopoly, all under the ownership and control of AT&T. This structure is shown in Figure 10.1. The most visible unit to the consumer was the local telephone company, which was responsible for operating and providing telephone service, including the leasing of telephone instruments, the wiring of premises, and the provision of dial tone and billing. The total package was bundled together as an end-to-end service with ownership of all facilities by the telephone company, as shown in Figure 10.2.

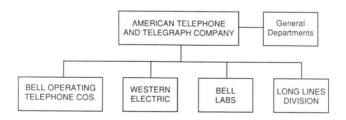

Figure 10.1 The old Bell System consisted of four major operating units, all under the control and direction of AT&T.

Twenty-two operating companies were in the Bell System and provided local telephone service. Many of them were corporate giants by themselves and were household names throughout the United States. New Jersey Bell, Southern Bell, Pacific Bell, Michigan Bell, New England Telephone, and Southwestern Bell were a few of the Bell operating companies in the old Bell System, the so-called BOCs.

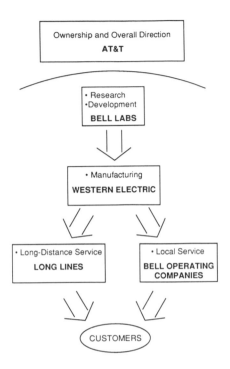

Figure 10.2 The Bell System was a vertically integrated monopoly covering all aspects of providing telephone service, including R&D, manufacturing, and the provision of local and long-distance telephone service.

Long-distance service between states is called *interstate toll service.* It was provided solely and transparently by the Long Lines Division of AT&T. A telephone subscriber making a long-distance call simply dialed the number and had no idea that Long Lines was involved in the call. Billing was handled by the local telephone company, and Long Lines was the only long-distance company in town—competition had yet to come to long-distance service. The local phone company gave all the money collected for long-distance calls to AT&T. To compensate the local companies for use of their local facilities in completing long-distance calls, AT&T returned more than half of the collected monies to the local telephone companies—a process called *separations.* The local companies relied on this source of revenue to keep the rates for local service low and affordable.

The transmission and switching equipment needed by the BOCs and by Long Lines was manufactured for them on a sole-source basis by the Western Electric Company. The trusted telephone instrument, which steadfastly survived falls on a hard floor, angry slammings of the handset, and spilled cups of coffee, was also manufactured by Western Electric and sold to the telephone companies for installation in the homes and offices of their customers. Western also made such items as the wire used in the provision of telephone service, the various electric components used in transmission and switching systems, and the batteries used to power the telephone system. Western made virtually everything, and the few items it did not wish to make were manufactured to Western specifications by a small cadre of chosen subcontractors.

All the items manufactured by Western Electric were developed and designed by engineers at Bell Telephone Laboratories. Much of the basic technology behind these designs flowed from research performed at Bell Labs. The quality of this research led to the high regard with which Bell Labs was held in the scientific and technological communities.

The stock of the phone companies and Western Electric was owned and held by AT&T. Bell Labs was owned jointly by Western Electric and AT&T, supposedly to acknowledge the relevance of much of the work at Bell Labs to Western's manufacturing mission. AT&T was thus a holding company, but its role extended far beyond being only a silent owner. AT&T was the "conductor" ensuring that the various "players" in the Bell "orchestra" all played together harmoniously. This leadership and direction role was performed by AT&T's General Departments and by AT&T senior management.

The AT&T General Departments addressed problems, issues, and questions common to all the telephone companies in providing telephone service. The work peformed by the General Departments was paid for by the Bell operating companies through a levy on their earnings, called the *license contract fee.* Ultimately, like all costs of doing business, the license contract fees were paid by the telephone subscribers. The General Departments also gave direction to Western Electric, and thus telephone subscribers were, in effect, helping to subsidize Western Electric. This became an issue as Western's competitors entered manufacturing and equipment provision.

The AT&T empire extended beyond telephone service. Nassau Smelting and Refining reclaimed precious and useful metals from scrap provided by the telephone companies from obsolete equipment and facilities. Sandia Laboratories developed nuclear weapons under contract to the federal government.

COMPETITION COMES TO TELECOMMUNICATION

The management of the Bell System came to believe that AT&T had a government-endorsed right to monopolize telecommunication in the United States. Perhaps this belief was well founded because on the many occasions that the government attempted to break up the Bell System, the various concessions made by AT&T were able to keep the Bell System intact and the government always abandoned its cases.

A serious attempt to dismantle the Bell System was initiated in 1949 by the Justice Department. That action was terminated in 1956 by the terms of the Final Judgment. The next attempt to dismantle the Bell System came 20 years later in 1974. That action was settled in 1982 under the terms of a modification of the 1956 Final Judgment. This Modification of Final Judgment finally resulted in the dismantling of the Bell System that the government had sought to achieve for nearly a century. Under the terms of the MFJ, the Bell operating companies were divested from AT&T; AT&T retained Western Electric and its Long Lines Division. This divestiture occurred on January 1, 1984. The Bell System was no more.

Such a drastic action was the result of many events and threats, but perhaps one factor overshadowed all the various events and threats. That factor was an attitude problem on the part of AT&T management. For decades, Bell System management believed that service to the public was the reason for its existence and that they alone knew best what was in the public's best interest. Management made decisions about what services would be overpriced in order to subsidize other services, supposedly in the best interest of the public. This type of thinking led to a corporate arrogance and intransigence that did not understand and was not open-minded to any changes or lessening of AT&T's monopolistic control of telecommunication in the United States. Thus, even the smallest threat to the *status quo* of the AT&T monopoly was met by an overwhelming, defensive response from the Bell System—and, indeed, a whole series of such threats occurred.

The first threat to the Bell monopoly, which proved to be relevant to its ultimate fragmentation, came in the 1960s from a small mobile radio communication company in Texas operated by the Carter Electronics Corporation. Carter's customers wanted to use their two-way radio equipment to connect with the telephone network to complete telephone conversations. Carter made an acoustic connection between a telephone handset and the radio equipment by using a device called a Carterphone. The phone company objected, citing the prohibition against any "foreign attachment" to its lines or equipment. At this time, even attaching a piece of cellophane tape to a telephone could bring the wrath of the phone company in response to this foreign attachment to its property. Carter was doing far more than that.

In 1968, the Federal Communications Commission decided that telephone tariffs barring connection of customer-owned equipment were unlawful. This decision initiated a series of telephone company actions aimed at impeding and delaying direct electrical connection to their lines. For a while, any direct electrical connection had to be made through the use of a protective-connecting arrangement (PCA) provided by the telephone company. In 1974, customer-owned equipment was allowed to be connected directly only if it contained special protective circuitry designed by Bell Labs. The rationale for these actions by the telephone companies was the claim that any direct connection might result in harm to the network and the telephone companies had a mission to protect the integrity of the network. I doubt if there is any real documented case of any such harm having ever occurred as a result of direct connection of any customer-owned equipment.

In 1977, the FCC attempted to rectify the situation. Customers were buying telephone answering machines and telephones and were connecting them directly to telephone company lines without any protective circuits. The FCC created a registration program in which the manufacturers of the equipment were required to show that their equipment met certain minimal standards. A certification number was then issued by the FCC.

The problem then became that the telephone company could no longer tell whether telephones connected to the line belonged to the phone company or to the customer. The telephone company began to require customers to call the company whenever the customer connected telephone equipment to the line and to report the FCC certification number of the equipment. The certification number contained information about the amount of current drawn by the equipment, and the phone company monitored the total current to determine what was connected to the line. The phone company even attempted for a while to collect a monthly line extension charge for the customer-owned equipment connected to the line.

Today, all this is quite different. The telephone companies are barred from providing telephone instruments as part of telephone service. Any customer-owned equipment can be connected to the line, although it must be type-certified by the FCC. This was facilitated by the development of the *modular plug and socket* by Bell Labs' engineers to make it easier for telephone company installers to connect and disconnect telephones. The same modular plug made telephones as easy to install as electric appliances—all you need do is simply plug them in. The Bell monopoly on telephones had finally died.

Another attack on the Bell System came in the 1960s involving the provision of long-distance private line service. The route between Chicago and St. Louis had a great amount of business service. AT&T's rates for this route, like other high-usage routes, were much higher than actual costs because long-distance subsidized local service and also rates were averaged across the nation. A competitor saw the opportunity to install its own private microwave towers between these two cities and to sell service in direct competition with AT&T. In 1969, the FCC authorized the provision of this private service by Microwave Communications Incorporated, or MCI as the company is known today.

Because private line service is a specialized offering, as opposed to the offering of conventional, public long-distance service, the companies offering such private line services were called *specialized common carriers*. Their existence was strongly opposed by AT&T, and a lengthy series of FCC and court actions ensued. The whole topic really exploded when MCI attempted to puchase connections to local telephone company lines to obtain access to the public switched network. In this way, subscribers to the MCI service could make a local call in Chicago to an MCI number, dial the desired number in St. Louis, connect via MCI facilities, and then obtain final connection over telephone company lines in St. Louis. MCI called this "new" service *Execunet.* AT&T's public long-distance service was clearly under attack. I emphasize again that the only ways in which MCI was able to compete with AT&T were (1) that MCI did not subsidize local service by paying large fees to the local telephone companies for access to the local network and (2) that MCI did not have to provide service to less lucrative cities and areas. AT&T's arguments against MCI were legitimate, but AT&T was viewed as a "nasty monopoly" simply protecting its position. AT&T turned to the political arena to obtain relief.

The Consumer Communications Reform Act (CCRA) was written by AT&T and introduced into Congress in 1976. A massive lobbying effort was then initiated by AT&T, urging employees and shareowners to write to their elected representatives to vote for the proposed legislation. The CCRA would have wiped out all competition and would virtually have returned to the pre-Carterphone days of the monopolistic past. Congress knows all about lobbying and recognized the CCRA for what it was. However, the result of the introduction of the CCRA was that Congress took a more active interest in telecommunication policy, and legislation favoring competition started to appear. AT&T then found itself actively lobbying to extricate Congress from telecommunication policy. I view the Congress as a hornet's nest, and hornets are best left alone. AT&T never should have introduced the CCRA; it was a serious blunder and indicated to me how panicked AT&T's management had become at that time.

In 1980, the FCC issued the results of its second investigation into computers and telecommunication. In *Computer Inquiry II,* the FCC stated that all telephone terminal equipment had to be detariffed by early 1982. Furthermore, AT&T was required to separate the provision of regulated services from that of deregulated services. Regulated services were called *basic,* and deregulated services were called *enhanced.* "Plain old telephone service," or POTS, is a basic service. Enhanced service pertains to applications and equipment that entail a resale value component, and this is also denoted as value-added service. Confusion between the two types of services still abounds.

The FCC clearly was attempting to promote competition in telecommunication and to force AT&T to reorganize, possibly divesting its manufacturing business. The FCC really did not have the authority to force AT&T to reorganize or to divest any business units, but as AT&T was continuing to lose most court cases, the FCC grew more bold in its dealings with AT&T.

THE COLLAPSE OF THE BELL SYSTEM

AT&T entered the early 1980s as a besieged company with many battle wounds and a very uncertain future. On one hand, competitors were allowed to do whatever they wished, but AT&T was still subject to regulation. And to make matters worse, the Justice Department antitrust case was still continuing in Judge Harold Greene's court with more uncertainty about its final outcome.

AT&T's management then saw a vision. That vision was an AT&T free to enter new businesses and markets based on the scientifc and technological inventiveness of Bell Labs. That vision saw the coming of the information age and an AT&T free to manufacture and provide computers and other information-age products and services. That vision saw plain old telephone service as being mundane and the way of the past. Furthermore, the local telephone companies were being viewed as losers with a declining and costly business. Concern over what might happen in Judge Greene's court, uncertainty about the future, and a new vision of the future motivated AT&T's management to seek a compromise settlement to the antitrust case. The settlement was made outside of Judge Greene's court to avoid the oversight provisions of the Tunney Act. Thus, AT&T and the Justice Department decided to reach a settlement that would be entered in the Newark court as a modification of the 1956 Final Judgment. This ruse to avoid Judge Greene's jurisdiction failed, and Judge Greene examined the settlement and finally accepted it in his court.

The settlement reached by AT&T and the Justice Department freed AT&T from all the restrictions of the 1956 Final Judgment. In return, the local telephone companies were divested from AT&T and were barred from manufacturing and from long-distance service. Intrastate toll service was given to AT&T. In terms of its new vision of the future, AT&T clearly won. Furthermore, AT&T had nothing to fear from any competition from the local companies, since they were barred from manufacturing and long-distance service—the "new" AT&T's two major business areas. It was almost as if AT&T itself wrote the modification of the Final Judgment to its best interests.

A large component of AT&T's business is the provision of long-distance service. Although AT&T had hoped otherwise, this business is still subject to regulation, and a host of competitors competes vigorously with AT&T.

There is an interesting way to view divestiture. In 1988, the total value of property, plant, and equipment (including accumulated depreciation) owned by the seven RHCs was $186.4 billion versus AT&T's total value of $40.9 billion. Clearly, the operating companies formed the bulk of the Bell System's assets, and the BOCs were the Bell System. Viewed this way, AT&T in effect divested itself of the Bell System!

Divestiture was a complex process, and it has taken some time for all the dust to settle. We shall see later in this chapter who really won—and lost—from the drastic restructuring of the telecommunication industry in the United States.

POSTDIVESTITURE BOC ORGANIZATION

As a result of divestiture, the Bell operating companies are all independent of AT&T (see Figure 10.3). The stock of the operating companies formerly owned by AT&T was divested and given to seven regional holding companies, which now

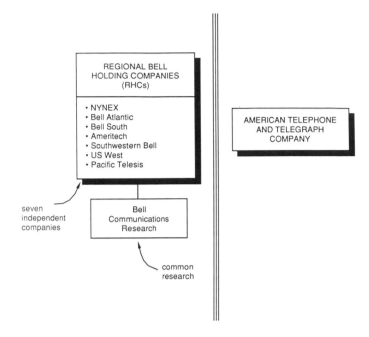

Figure 10.3 As a result of divestiture, AT&T and the Bell operating companies are totally independent of each other. The stock of the BOCs is owned by seven regional holding companies. Bellcore is owned by and performs common research and other activities for the seven RHCs.

own the BOCs. The seven regional holding companies and the BOCs that comprise them are:

- Ameritech
    Illinois Bell
    Indiana Bell
    Michigan Bell
    Ohio Bell
    Wisconsin Bell
- Bell Atlantic
    Bell of Pennsylvania
    Chesapeake and Potomac Companies
    Diamond State
    New Jersey Bell
- BellSouth
    South Central Bell
    Southern Bell
- NYNEX
    New England Telephone
    New York Telephone

- Pacific Telesis
  Pacific Telephone
  Nevada Bell
- Southwestern Bell
  Southwestern Bell
- US WEST
  Mountain Bell
  Northwestern Bell
  Pacific Northwest Bell

AT&T identified the need for the BOCs to have their own central research organization to address common technical and scientific questions. This need was addressed by the creation of a central service organization, called Bellcore, fully owned and funded by the seven RHCs to be responsive to their needs and direction.

The provision of local service is the main business of the BOCs and thus their owners, the RHCs. Many of the RHCs have entered other businesses, as allowed by Judge Greene. However, most of the revenue and profits of the RHCs still come from the BOCs and the provision of local telephone service. We shall see later in this chapter that a large portion of that revenue comes mostly from AT&T in the form of access charges. AT&T and the BOCs are still pretty much welded together.

POSTDIVESTITURE AT&T ORGANIZATION

Since divestiture, AT&T has undergone a series of reorganizations as it attempts to redefine its business and markets. The frequency of the reorganizations along with their complexity make it a challenge to comprehend how and why AT&T is organized the way it is at any particular moment, but perhaps learning how to cope with uncertainty and constant change is the challenge of the day, not only for AT&T. Even the Bell logo has evolved with divestiture, as shown in Figure 10.4.

Immediately after divestiture, AT&T was organized into three major units. AT&T Communications provided long-distance service to business and residential customers. AT&T Information Services manufactured, sold, and leased telecommunication products to consumers, large business customers, small business customers, and retail outlets. AT&T Technologies manufactured and supplied telecommunication equipment, such as network switching and transmission systems along with various components ranging from transistors to power supplies. Most of the network equipment was sold to telephone companies. This organizational structure was similar to the old AT&T with AT&T Communications being Long Lines, with AT&T Technologies being Western Electric, and with AT&T Information Services managing the customer interface in a fashion akin to the telephone companies.

Figure 10.4 The Bell logo has evolved over time, but prior to divestiture, the "bell" was always an essential feature. After divestiture, AT&T introduced its new logo to indicate its new "world" view. (Photo courtesy of AT&T Bell Labs.)

This structure was abandoned in the mid-1980s for a number of separate business units, called groups, to serve customers and markets. The Data Systems Group marketed AT&T computers, local-area networks, and other data networking equipment to large and small business customers. The Network Operations Group operated the AT&T long-distance network and also installed telecommunication systems for large business customers. The Business Markets Group and the General Markets Group provided voice and data communication products and services to business and to residential customers, respectively. These four groups formed what was called internally "the end-user organization." The Network Systems Group manufactured, installed, and serviced the switching and transmission systems used by local telephone companies. AT&T Microelectronics belonged to the Network Systems Group and manufactured the various components used in AT&T products. The Technology Systems Group supplied products and services to the federal government and also managed the Sandia National Laboratories. AT&T Bell Labs supplied R&D services to all of the six groups.

In the late 1980s, a new organizational structure was created consisting of 18 strategic business units (SBUs) to serve AT&T customers. As examples, the Consumer Communications Services unit is responsible for basic long-distance service for consumers, and the Consumer Products unit is responsible for marketing telephones and other such products to the consumer market. Support to the SBUs is provided by 24 operating divisions. This type of organization appears to be a so-called matrix structure, with markets and products along one axis (the SBU axis) and operations along the other axis (the operating-division axis). As an example, the Network Services Division is responsible for operating AT&T's long-distance network and works for both the Consumer Communications Services unit and the Business Communications Services unit. AT&T Bell Labs is another operating division and provides R&D services to the various SBUs.

AT&T's organizational structure appears to be getting larger and more complex over time: 3 major units in the early 1980s became 6 major groups in the mid-1980s and then grew to a total of 42 business units and operating divisions in the late 1980s. The motivation for all these changes appears to be an attempt to be more responsive to customers and also to make it easier to determine bottom-line financials for various businesses. While these motivations are most certainly noble and worthwhile, the frequent reorganizing can create considerable confusion and frustration for both employees and customers. Also, the present matrix structure can cause much overlap of responsibility and many opportunities for diffused authority. The considerable growth in the number of organizational units experienced by AT&T can sometimes result when personalities and fiefdoms dictate organizational structure. Organizational theorists will continue to watch AT&T with great interest as its structure undoubtedly changes again and continues to evolve in the future.

ACCESS CHARGES

To provide long-distance service, carriers need access to the local network both to complete a long-distance call in the distant city and to give access to the calling party. Long-distance carriers pay the local telephone companies for this access through the imposition of access charges. In 1988, the RHCs received 27 percent of their revenue in the form of access charges paid by long-distance carriers, such as AT&T. In that same year, AT&T paid 47.5 percent of its long-distance receipts to local telephone companies for access to their local facilities. Access charges are the glue that continues to maintain an intimate financial relationship between AT&T and the BOCs.

Access charges are not new, although in the past they were called separations. Long Lines returned a large proportion of the charges collected by the local telephone companies back to them as reimbursement for the use of their local facilities in completing long-distance calls. Nearly 60 cents of every dollar in toll charges was returned to the local company and, in theory, helped subsidize local service. The local companies encountered costs in supplying billing, switching, common local transmission, and other services associated with long-distance service. These costs are difficult to determine, and thus I am not sure whether the 60 cents on the dollar was really sufficient payment to the local telephone company. However, the accepted wisdom was that toll service subsidized local service. The returned money was a sizable source of revenue for the local companies and perhaps could be viewed as the financial glue that kept the Bell companies in the AT&T empire.

With divestiture, the subsidy—if indeed it were one—still had to be paid or the local companies would lose a large source of income and local rates would need to be increased accordingly. The subsidy became an access charge, initially paid solely by long-distance carriers. To decrease the access charges paid by these carriers, the FCC imposed a flat increase in local rates to be paid by all telephone subscribers, and this new fee was also called an access charge. The rationale for the subscriber access charge was that if carriers were being charged too much for access they would bypass the facilities of the local telephone company and connect directly to large business customers. This would deny revenue to the local company, and local rates would increase.

One way to reduce the costs of access charges would be to allow the BOCs to provide intrastate long-distance service. The BOCs provided this service before divestiture, but the stipulations of the MFJ allowed AT&T to take away this very profitable business from the BOCs.

Access charges continue to be a controversial topic and are one example of the lack of foresight of all the consequences of competition in telecommunication and of divestiture.

THE RHCs

Prior to divestiture, the stock of the Bell operating companies was owned and held by AT&T. During the process of divestiture, AT&T created seven regional holding companies and transferred the ownership of the BOC stock to the RHCs. Stock in the RHCs was distributed to AT&T shareowners in proportion to their AT&T holdings. One rationale for the creation of the RHCs was AT&T's belief that some of the BOCs were too small to be financially viable. Combining BOCs would create seven larger and stronger companies. However, the regional holding company is simply a smaller replacement of the AT&T bureaucracy. The BOCs never achieved true independence—they now are owned by RHCs.

After divestiture, the RHCs attempted to enter new business areas. Judge Harold Greene, who presided over the divestiture case and continues to oversee its provisions, put a halt to some of these planned ventures and stressed that the major business of the RHCs is the provision of local telephone service within their assigned local access and transport areas. The fear is that the RHCs will subsidize the new business ventures from the profits of their regulated telephone business. The RHCs continue to hope for relief from these restrictions. In particular, some of the RHCs would like closer involvement with manufacturers, if not actual entry into manufacturing. Although, why anyone would want to be a manufacturer in today's information and service economy eludes me.

The antitrust case whose settlement resulted in divestiture was primarily against AT&T, Western Electric, and Bell Labs. Nevertheless, it was the BOCs that were "punished" by being divested from AT&T, and AT&T was "rewarded" by being freed from the restrictions of the 1956 case. The answer to who was really punished and who was really rewarded is best determined by examining the financial effects of divestiture on AT&T and the RHCs.

FINANCIAL EFFECTS OF DIVESTITURE

In 1982, the year before divestiture, AT&T had a net income of $7.3 billion, or a net income equal to 11.1 percent of revenue (commonly know as profit margin). In 1984, the first full year of operation after divestiture, AT&T's net income dropped to $1.4 billion with a profit margin of only 2.6 percent. In reporting its income after divestiture, AT&T routinely excludes the access charges it pays to the local telephone companies. Since these access charges are collected by AT&T and represent a cost of doing business, I have included them in my calculations of AT&T's revenues. In 1984, the RHCs as a group had a net income of $6.8 billion and a profit margin of 11.8 percent. Clearly, the BOCs were the financial strength of the Bell System, and AT&T was crippled financially as a result of their divestiture.

Figures 10.5 and 10.6 show graphically the financial results of divestiture in terms of net income and profit margin from 1982 to 1989. The dip in AT&T's financials in 1986 was caused by the setting aside of a large sum of money to finance early retirements and to stimulate resignations in order to reduce the size of AT&T's labor force. The 1988 loss was the result of accelerated depreciation and the replacement of analog multiplexing facilities in AT&T long-distance network. The trend, ignoring these two dips, in AT&T's financials was modest growth in net income at a compound annual growth rate of 14 percent over the five-year period from 1984 to 1989.

The financial effects of divestiture clearly show the considerable health of the BOCs and RHCs. The surprise was the comparatively poor performance of AT&T. We shall examine some possible reasons for this later.

RESEARCH AND DEVELOPMENT

In this book, we have seen the importance of science and technology and of research and development (R&D) in the provision of telephone service. Much of

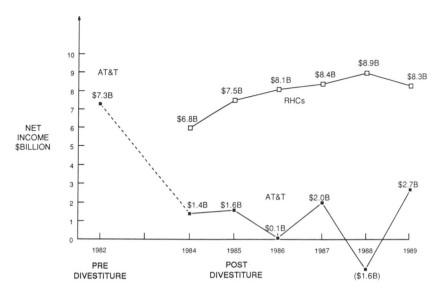

Figure 10.5 The net income of the seven regional holding companies has climbed steadily since divestiture. The small decrease in 1989 was a result of an earnings decrease at NYNEX. AT&T's earnings have varied greatly since divestiture. The 1986 dip was a result of putting aside funds to stimulate employees to resign or retire early. The 1988 dip was a result of accelerated digitization of the AT&T network.

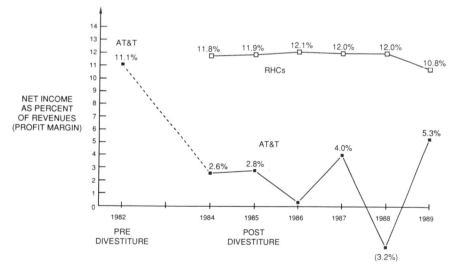

Figure 10.6 The profit margins of the seven regional holding companies and of AT&T are compared. The calculation of AT&T's profit margins includes access charges. Discounting dips in 1986 and 1988, AT&T's overall trend shows growth in profit margins, although the figures are at least half those of the seven RHCs.

that R&D was performed at Bell Laboratories. Bell Labs' engineers designed the equipment and systems that were manufactured by Western Electric and used by the Bell operating companies and Long Lines. About 10 percent of Bell Labs' resources were devoted to performing basic research in areas deemed applicable to telecommunication, and much of Bell Labs' reputation was based on the accomplishments of its researchers. Indeed, Bell Labs had come to be considered a national R&D resource. Quite understandably there was concern that divestiture and the restructuring of telecommunication might somehow damage Bell Labs and telecommunication R&D. I address that issue in this section.

A few years after divestiture, I performed a study to assess the effects of divestiture on Bell R&D. Two years before divestiture, at the end of 1981, AT&T spent $1.6 billion for R&D conducted mostly at Bell Labs, and 24,100 people were employed at Bell Labs. With divestiture, Bell Labs was required to change its name to indicate its continued affiliation with AT&T, and Bell Labs was renamed AT&T Bell Labs. A new organization, Bellcore, was created to serve the research, development, and other common needs of the divested BOCs and their RHCs. Hence, the postdivestiture R&D picture must combine the AT&T R&D effort with Bellcore. By doing this combining, I found that at the end of 1985 there were 32,000 R&D employees at a combined funding level of $3.2 billion. The number of researchers had likewise increased from a predivestiture number of 1200 to a postdivestiture combined number of 1700.

In terms of numbers and figures, R&D benefited as a result of divestiture. As we enter the 1990s, new questions about the health of telecommunication R&D must be raised. For one, it appears that for 1989 AT&T spent nearly $2 billion on product-related R&D. Actual sales of products generated only $11.5 billion, and hence product R&D was more than 17 percent of sales. This is very high and indicates the high cost of AT&T R&D. We shall see in the next section that this R&D is, in effect, subsidized by the profits generated by AT&T's very successful long-distance business.

At the opposite end of the picture, in 1989 the RHCs spent about $1 billion for R&D performed for them by Bellcore. This amounts to less than 1.5 percent of their combined revenues of $77 billion. Given the high-technology and complexity of today's telecommunication system coupled with its essential role in our information and service economy, I wonder whether this amount of R&D is adequate. Since the RHCs are forbidden from any manufacturing activities, however, perhaps much less R&D is needed for their predominantly service-based network business.

Bell Labs of the past was a great research organization. I know this firsthand since I worked there for well over a decade performing research in a number of exciting and novel areas. What contributed to the greatness of Bell Labs as a research institution were such factors as secure, long-term funding; a strong sense of a mission to ensure the long-term future of telecommunication; and the freedom to take risks and to fail. AT&T Bell Labs is not Bell Labs of the past for the simple reason that its owner, AT&T, is not the same as predivestiture AT&T. AT&T is now subject to competition and revenue fluctuations; secure, long-term funding cannot be assured anymore. AT&T's mission changes with management whims; a strong sense of mission is absent. AT&T must be concerned about new products and services and their success in the marketplace; the freedom to take risks is much reduced. The new emphasis is on relevance; research must be more applied and its results must be quickly transferred to technology relevant to existing AT&T products, services, and businesses. Yet technology transfer is a tricky business and frequently occurs outside formal organizations.

The health and state of research at AT&T Bell Labs really is AT&T's concern. However, the health and state of research in telecommunication in general should be the nation's concern. Since it was AT&T that claimed before divestiture that Bell Labs was a national treasure and resource, the health of postdivestiture research at AT&T Bell Labs continues to be a national question. The real issue though should not be AT&T Bell Labs in particular, but rather research in telecommunication in general. The present situation in which AT&T has been funding research performed at AT&T Bell Labs that benefits the whole telecommunication industry is untenable in the long run. Somehow, the whole industry must assume more responsibility for funding research. I do not know the solution, but perhaps some combination of the research performed at Bellcore and at AT&T Bell Labs would make sense with broad funding from the whole industry.

AT&T: A STATUS REPORT

AT&T has three major businesses: sale of services, in particular, telecommunication services such as long-distance telephony; sale of products, such as telephones, switching systems, and PBXs; and rentals. In 1989, these three business segments generated $22.1 billion (net of access charges of $14.9 billion), $11.5 billion, and $2.5 billion, respectively. AT&T does not report the actual profit or loss for each separate business, but by examining the data reported to the FCC by AT&T for its communication service business and by making some assumptions, some figures can be extracted. These figures show that AT&T's product business could have lost as much as $4 billion in 1989. AT&T's service business has become the "cash cow" that keeps AT&T afloat, but just how much longer can that cow be milked?

Given the size of these losses, why would AT&T retain the sale of products? A number of possible explanations are possible. One explanation is that my estimations of the losses are incorrect, but I do not think so. Another explanation may be that AT&T hopes to be able to turn the product business around and make it profitable in the future. Another explanation may be that AT&T believes it must be able to provide products as part of a total system solution to the communication needs of its customers. A last reason may be that AT&T needs to shelter the profits of its long-distance business or else it may be subject to greater government scrutiny and regulation.

From a purely financial perspective, AT&T should divest or otherwise eliminate the sale of products. The present organizational structure with its many strategic business units might make it relatively easy to sell off individual product units. If AT&T continues to retain its product business, the cost of product-related research and development needs to be reduced by about $1 billion and corporate overhead (selling and administrative costs) needs to be reduced by about one-third. These two drastic actions would save about $4 billion and would offset the losses in the product business. AT&T's matrix organization is also very costly and results in much duplication and overhead; it needs an overhaul in the direction of simplification and consolidation.

If AT&T were to divest or sell its entire manufacturing and sale-of-products business, a much leaner and more sharply focused AT&T would result. It is possible that the government might be willing to give AT&T relief from regulation in return. The AT&T service business would be very free to purchase its equipment and systems from any one of many suppliers. This would stimulate the equipment business in a way similar to what happened when the BOCs were freed by divestiture from purchasing most of their equipment from AT&T. Research at AT&T Bell Labs would also benefit since it would have a source of stable, long-term funding with a more focused and clearly stated mission.

AT&T has great economies of scale in the provision of long-distance and other telecommunication services. Hence, AT&T's service business is very profitable. However, the large losses in the sale of products undermine the overall profitability of AT&T. If these losses were eliminated through divestiture or other elimination of the product business, AT&T's long-distance rates could be reduced thereby forcing AT&T's competitors into financial jeopardy. I have always believed that competition in long-distance telephone service does not make sense, and I believe that if AT&T did not have to subsidize the losses from its product business, it would eliminate all competition by providing superior service at the lowest price to the consumer. How would the federal regulators react to this return to a near monopoloy situtation in the provision of telecommunication service?

There certainly appears to be much controversy and opinion about AT&T's future. Most certainly this future will not be stable and much change is probably still in store for AT&T.

THE BOCs: A STATUS REPORT

On the whole, the BOCs have been a great success. Their profit margins are stable and quite good. Productivity measured as the number of access lines per employee continues to grow. In 1990, well over 90 percent of their access lines was served by computer-controlled switching systems. Network usage continues to grow. By any and all measures, the BOCs have been a great success. And yet, the BOCs are not content with what they have.

The BOCs want to be able to provide information services and to be involved with manufacturing. In the late 1980s, the BOCs initiated advertising and lobbying efforts to obtain relief from the restrictions of the Modified Final Judgment so that they could enter these other businesses. So far, the advertising effort seems to have missed the mark. A student of mine performed an evaluation of one of the advertisements and found that most people had no idea what the ad was about and the few that did easily saw through it. The lobbying effort has failed too. Like AT&T's misguided Consumer Communications Reform Act, the BOCs have attracted the attention of Congress, and who knows where, if anywhere, that will lead.

To some extent, the BOCs have good reason to want to be involved with manufacturing. Closer ties with manufacturers might allow the BOCs to help speed the innovation and introduction of new telecommunication services and products. However, given AT&T's problems with manufacturing and the sale of products, the BOCs would be well advised to avoid becoming an actual manufacturer. The restrictions of the MFJ against manufacturing have saved the BOCs from replicating AT&T's losses in the product area.

The BOCs still provide telephone directories along with their very profitable yellow pages listings and advertising. Won't it be nice to have a shopping service in which you telephone a directory operator who then uses a computer terminal to search a database and gives you the name and address of the nearest stores along with information about any special sales? Should such a service be supplied by the telephone company or by others? Should the telephone company be forbidden, as it now is, from supplying such a service? A laser disk can easily store all the information in a number of telephone books. Should the telephone company create and distribute such disks? Some people still believe that computerized databases for the home—videotex—are the way of the future. Should the telephone companies be allowed to own and operate such videotex services? What about owning and operating gateways to databases owned and operated by others?

The issues and questions concerning the provision of information services by the BOCs and RHCs are most challenging and interesting. One thing for certain is that many of these new information services are very risky and could be a great opportunity to lose many millions—perhaps even billions—of dollars. The one business that is not risky is the transmission and switching of the signals that give access to these services. The transmission and switching of signals—the conduit business—is indeed the business of the BOCs. In most cases, the BOCs would probably be well advised to avoid the hype and risk of many of these new information services and instead concentrate on the transmission and switching of signals.

A FINAL ASSESSMENT

It is indeed difficult to gaze into a crystal ball and ascertain the ultimate effects of divestiture on telecommunication in the United States. The 1980s saw a great upheaval in telecommunication, and I wonder what surprises are in store for the 1990s. Will AT&T divest its manufacturing and product business? Will the BOCs be allowed to enter manufacturing, and if so, will they fare any better than their past owner? Will the BOCs enter the long-distance business? What will be the performance of Northern Telecom, MCI, Sprint, and all the other firms in the telecommunication industry?

In gazing into the crystal ball in the first edition of this book, I raised the question whether AT&T would become an entrepreneurial company able to develop and manage a broad portfolio of technology businesses. Clearly, that has not happened. Competition with AT&T's products has been very keen, and AT&T's foray into developing its own personal computer was a failure and financial disaster.

There was a tendency in the past to assume that "telecommunication" was synonymous with AT&T and the Bell System. Divestiture most certainly changed that assumption. There are a number of long-distance carriers from which to choose, although AT&T clearly remains the dominant carrier. The BOCs still have a virtual monopoly on local service, but bypass and the use of new technologies are mentioned as possible future threats to their dominant position. Telecommunication has become a partnership involving a variety of competitive firms, which together provide end-to-end service. The challenge is to be sure that this sum of parts continues to equal the quality of service of the past.

Competition in customer-premises equipment has resulted in higher quality telephones at lower prices to the consumer. Competition in long-distance service has resulted in considerable savings to those making such calls. However, the rates for local service have increased greatly. This increase was partially an overreaction to the uncertainties of divestiture and also the need to accelerate the depreciation and replacement of old electromechanical switching equipment. Rates for local service should stabilize and actually should decrease as the efficiencies of the newer switching equipment take effect. Overall, it would appear that the consumer has benefited from competition in telecommunication and from the divestiture of the BOCs from AT&T.

Compared with the rest of the world, the United States has excellent telephone service at the lowest rates. Telephone service in most of the world is still a government monopoly, but that situation is changing as some countries follow the United States in seeking more competition in telecommunication and by privatizing its provision.

The major problem with the present competitive system is that no single supplier is responsible for the total end-to-end provision of telephone service. Decisions must now be made about such items as which company to obtain long-distance service from and whose telephone to purchase. Decisions are hard to make, and thus far, that seems to be the major cost of divestiture and competition.

Clearly, the dust from the explosive turmoil of divestiture and the restructuring of telecommunication in the United States is still in the process of settling. Uncertainties, challenging issues, and intriguing questions abound. It certainly seems that the last decade of this century will see many more changes in the structure of the telecommunication industry in the United States. The "last chapter" of the story of restructuring has not yet been written.

THE LAST CHAPTER

Our book has told the story of the technology of telephone systems and telecommunication. The technology is indeed a marvel and has greatly benefited us in terms of low cost and useful features. Telephony and telecommunication are the key to the information age and our service economy. We may take telephone service for granted, but we would be lost without it and our planet would come to a halt—it would be *The Day the Earth Stood Still.*

Telephony and telecommunication is a system, and we have learned in our book how station apparatus, transmission, switching, and signaling work together as that system. However, there are other parts of the system that we have only touched on. People and institutions, too, are a part of the system that brings us telephone service. Scientists and engineers contribute the science and technology of telephony. Managers and workers make the service work for us from day to day. And the wondrous visions of Bell and Vail, and others, continue to illuminate the way of the future.

REFERENCES

Noll, A. Michael, "Bell System R&D Activities: The Impact of Divestiture," *Telecommunications Policy,* Vol. 11, No. 2, June 1987, pp. 161–178.

Noll, A. Michael, "The Effects of Divestiture on Telecommunications Research," *J. of Commun.,* Vol. 37, No. 1, Winter 1987, pp. 73–80.

GLOSSARY

This glossary defines terms and concepts that are used in the book but are not defined as used.

amplifier—an electronic device that makes a signal larger in amplitude. The *gain* of an amplifier is a measure of how much bigger the output is compared to the input. The gain is usually measured in *decibels,* or *dB.*

amplitude modulation (AM)—a technique for varying the maximum amplitude of a high-frequency tone, or *sine wave,* called the **carrier,** in synchrony with the instantaneous variations of the amplitude of an input signal, such as speech or music. The effect of amplitude modulation is to translate, or shift, the spectrum of the input signal to a new frequency range. The translated spectrum consists of two portions, called **sidebands.** One is an exact replica of the spectrum of the input signal and is located above the carrier frequency. This portion is called the **upper sideband.** The second portion is located below the carrier frequency and is an exact mirror image of the spectrum of the input signal; it is called the **lower sideband.** Because the same information is contained in both sidebands, sometimes only a single sideband is transmitted—a technique called **single sideband** (SSB) transmission. Also, the carrier is sometimes suppressed because it too conveys no information; this is called **suppressed carrier** (SC) transmission.

bandwidth—the width of the frequencies that comprises a signal or that is allowed to pass through a communication channel. According to the French mathematician Fourier, any complex waveform can be decomposed into the sum of pure tones, or *sine waves,* at different frequencies and with different maximum amplitudes and phases. A graphical plot of the frequencies that comprise a signal, or that are passed by a communication channel, is called the **spectrum.**

bps—the abbreviation for **bits per second.** A **bit** respresents a single *binary digit* and is either a zero or a one.

capacitor—an element used in electrical circuitry, consisting of two conducting surfaces, or plates, separated by an electric insulator. A capacitor blocks the flow of direct current (dc) but allows alternating, or changing, current (ac) to flow. The higher the frequency of the alternating current, the less opposition encountered to its flow across the capacitor. A capacitor has a **capacitance** associated with it, according to its ability to block lower frequencies. Capacitance is measured in the unit of farads (F).

current—a measure of the number of electrons flowing per second in an electric circuit. The unit of current is the **ampere,** abbreviated A.

dB—the abbreviation for **decibel,** which is a measure of comparison between two quantities. The decibel is the logarithm of the ratio of the two quantities. If the quantities being compared are electric powers, then a change of 3 dB means that one power is twice as large as the other.

dc—the abbreviation for **direct current.** Direct current is a constant, unchanging electric current that does not vary with time and that flows in one direction. A battery generates a direct current.

diode—an element used in electrical circuitry that passes current flowing in one direction but blocks current attempting to flow in the opposite direction. Diodes are usually constructed from a junction of semiconducting material.

duplex (full, half)—a form of two-way communication. With **half-duplex** communication, only one party can communicate in one direction at a time. With **full-duplex** communication, each party can communicate simultaneously in each direction. The push-to-talk system of two-way radio communication is an example of a half-duplex communication system. The telephone system is mostly full-duplex. Communication that is strictly one way, such as broadcast television, is called **simplex** communication.

filter—electrical and electronic circuitry designed deliberately to pass or to block certain frequency components in a signal. A filter that passes high frequencies is called a **high-pass filter** (HPF). A filter that passes only low frequencies is called a **low-pass filter** (LPF).

frequency modulation (FM)—a technique for continuously changing, or *modulating,* the frequency of a high-frequency tone, or *sine wave,* in synchrony with the instantaneous variations in amplitude of an input signal, such as speech or music. Frequency modulation translates and also expands the spectrum of the input signal. An FM signal has greater *noise immunity* than an AM signal, but this increased noise immunity is obtained at the expense of greater bandwidth.

frequency shift keying (FSK)—a technique for transmitting data in which the frequency of the carrier wave is changed from one value for a binary zero to another value for a binary one. Frequency shift keying is a form of frequency modulation. The term *keying* comes from its initial use in the days of telegraph keys.

Hz—the abbreviation for **Hertz,** which is a unit of measurement of the **frequency** of a time-varying waveform. The frequency of a *periodic waveform* that has a basic shape that continuously repeats is the number of full repetitions, or *cycles,* that occurs in one second. High frequencies are usually written using Greek abbreviations. A frequency of 1000 Hz is written as 1 kHz, where "k" stands for *kilo.* A frequency of 1,000,000 is written as 1 MHz, where "M" stands for *mega.* A frequency of 1,000,000,000 is written as 1 GHz, where "G" stands for *giga.*

impedance—the opposition to the time-varying current in an electric circuit. Impedance is measured in the units of ohms (Ω).

inductance—a measure of the ability of a coil of wire, called an **inductor,** to block high-frequency signals from flowing through it. Inductance is measured in the units of henrys (H).

mA—the abbreviation for *milliamperes.* An **ampere** is a measure of the current flowing in an electric circuit. Current is a measure of the number of electrons flowing per second in the circuit.

ohm—the unit of measurement for the opposition to the flow of current in an electric circuit. The ohm is denoted by the Greek letter omega (Ω).

Ohm's law—the relationship between the electromotive force E, electric current I, and resistance R in an electric circuit. Ohm's law is $E = IR$. This equation tells us that the electric current increases proportionately with the electromotive force.

phase-shift keying (PSK)—a technique for transmitting data signals in which the phase of the carrier wave indicates whether a binary zero or a binary one is being sent. With **differential phase-shift keying (DPSK),** a change of phase indicates the binary information.

resistance—a measure of the opposition to the flow of direct current in an electric circuit. Resistance is measured in ohms (Ω).

rms—the abbreviation for **root mean square.** The rms value of a time-varying waveform is the effective value of the waveform in terms of a constant, nonvarying waveform that would generate the same amount of energy. The rms value is calculated as the square root of the mean value of the waveform squared at each instant of time.

sine wave—a pure tone consisting of a single frequency. A sine wave is characterized by its maximum amplitude, frequency, and phase. The phase is a measure of when the sine wave starts. A sine wave is a smoothly time-varying waveform that alternates in *polarity,* or direction of flow (either positive or negative).

transformer—an electric device consisting of two coils of wire that are closely coupled magnetically with each other. A changing signal flowing in the input coil, or **primary,** induces a similarly shaped signal in the output coil, or **secondary.** The voltage amplitude of the output signal can be made larger or smaller than the input signal, depending on the ratio of the number of turns of the primary compared to the secondary. Sometimes there are two secondary coils obtained by connecting a wire, called a **tap,** at the center of the secondary coil; this is called a **center-tapped** secondary.

volts—a measure of the electron-moving, or electromotive, force (emf) that causes electric current to flow in an electric circuit. Volts are abbreviated V.

watts—a measure of the electric power of a signal, abbreviated W.

BIBLIOGRAPHY

Blyth, W. John, and Mary M. Blyth, *Telecommunications: Concepts, Development, and Management,* Indianapolis, IN: Bobbs-Merrill, 1985.

Cannon, Don L., and Gerald Luecki, *Understanding Communications,* Dallas, TX: Texas Instruments, Inc., 1980.

Engineering and Operations in the Bell System, 2nd Ed., Murray Hill, NJ: AT&T Bell Laboratories, 1983.

Fihe, John L., and George E. Friend, *Understanding Telephone Electronics,* Dallas, TX: Texas Instruments, Inc., 1983.

Freeman, Roger L., *Telecommunication System Engineering,* New York: John Wiley and Sons, 1980.

Gurrie, Michael L., and Patrick J. O'Connor, *Voice/Data Telecommunications Systems,* Englewood Cliffs, NJ: Prentice-Hall,1986.

Martin, James, *Future Developments in Telecommunications,* 2nd Ed., Englewood Cliffs, NJ: Prentice-Hall, Inc. 1977.

Martin, James, *Telecommunications and the Computer,* 2nd Ed., Englewood Cliffs, NJ: Prentice-Hall, Inc., 1976.

Parker, Sybil P., ed., *Communications Source Book,* New York: McGraw-Hill Book Company, 1987.

Pierce, John R., and A. Michael Noll, *Signals: The Science of Telecommunications,* New York: Scientific American Library, 1990.

INDEX

A band, in cellular telephony, 148, 153
A-law companding, 38
A-type channel bank, 35
Access charges, 54, 218
ACD (automatic call distribution), 134
Address
 in data communication, 163–164
 in signaling, 140
Administrative module (AM), in 5ESS
 machine, 127
Advanced mobile phone service (AMPS), 148
Airfone® system, 199
Alerter, in telephone, 29
Alerting function, in signaling, 140
ALI (automatic location information), in 911
 service, 202
Alias frequencies, 37
American Bell Telephone Company, 11, 178
American Speaking Telephone Company, 178
Ampere, unit of current, 230
Amplifier, 229
Amplitude modulation (AM), 39, 229
AMPS (advanced mobile phone service), 148
Analog multiplexing, 35
Analog *versus* digital, 143
ANI (automatic number identification), in 911
 service, 202
Anti-sidetone circuit, 21
Antitinkling filter, in telephone, 27
Antitrust actions, 182, 184
Application layer, 165–166
Appollo® airline reservation system, 194
AR6A system, 46
Area codes, 145
ARPANET, 166

ASCII, 162
Asynchronous data communication, 163
AT&T (American Telephone and Telegraph
 Company), 4, 208
 chartering of, 180
 digital radio systems, 47
 General Departments, 210
 picturephone service, 188–190
 postdivestiture organization, 214–216
 status report, 223
 switching systems, 109–119, 127–132
AT&T Bell Labs, 222–223
Attenuation, 34
Audiotex, 196
Automatic call distribution (ACD), 134
Automatic Electric Company, 14, 103
Avalanche diodes, 65

B band, in cellular telephony, 148, 153
B channel, in ISDN, 170
Balance network
 in network hybrid, 74
 in telephone, 22
Bandwidth, defined, 229
Banking, electronic, 1
Base station, in cellular telephony, 151
Baseband digital data, in LANs, 169
Baseband signal, 36
Basic rate interface, in ISDN, 170
Baud, 47
Bell, Alexander Graham, 7, 35, 61, 177,
 207
Bell operating companies (BOCs), 185, 208,
 214–216
 status report, 224–225

235

loose contact, 11
variable resistance principle, 7
Transoceanic undersea cable, 70
Transponders, 50
Transport layer, 165–166
Traveling-wave tube (TWT), 56
Trunk-busy signal, 141
Trunk-link network, in No. 5 crossbar, 112
Trunks, 4, 81, 139
TSIU (*see* Time-slot interchange unit)
TST switch example, 99
Tunney Act, 185
Twisted pair, 31
Two-wire-to–four-wire conversion,
   in network, 74–76
   in telephone, 22
TWT (traveling wave tube), 56

Undersea cable, 70
Universal telephone service, 207
Uplink, in satellite systems, 51
Utopian view of communication technology, 1

Vail, Alfred, 179
Vail, Theodore N., 5, 178, 207
Variable-resistance transmitter, 7

Varistor, in telephone, 27
Vertically integrated monopoly, 208
Video teleconferencing, 1, 190–191
Videophone, 79, 188–190
Videotex, 5, 191–195
VideoWindow® teleconferencing system, 191
Viewdata, 193
Viewtron videotex service, 193
Vocoders, 174
Voice-band interface unit, in No. 4 ESS
   machine, 119
Voice mail, 173
Volt, unit of electromotive force, 232

Watson, Thomas A., 5, 7, 12
Watt, unit power, 232
Waveguides, 42
Western Electric, 177, 210
   formation of, 179
Western Union, 178
Westrex, 183
Wheeler, William, 60
White noise, 78
Wickersham, George W., 181
Wire, 31–32

The Artech House Telecommunications Library

Vinton G. Cerf, *Series Editor*